# Undead Science

# Undead Science

## *Science Studies and the Afterlife of Cold Fusion*

BART SIMON

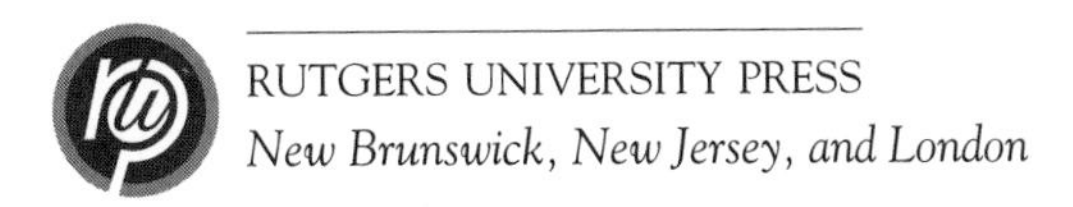

RUTGERS UNIVERSITY PRESS
*New Brunswick, New Jersey, and London*

**Library of Congress Cataloging-in-Publication Data**

Simon, Bart, 1966–
Undead science : science studies and the afterlife of cold fusion/Bart Simon.
p. cm.
Includes bibliographical references and index.
ISBN 0-8135-3153-5 (cloth: alk. paper)—ISBN 0-8135-3154-3 (pbk.: alk. paper)
1. Cold fusion. 2. Science—Social aspects. I. Title.

QC791.775.C64 S56 2002
306.4'5—dc21

2002023701

British Cataloging-in-Publication information is available from the British Library.



Manufactured in the United States of America

*True!—nervous—very, very dreadfully nervous I had been and am; but why* will *you say that I am mad? The disease had sharpened my senses—not destroyed—not dulled them. Above all was the sense of hearing acute. I heard all things in the heaven and in the earth. I heard many things in hell. How, then, am I mad? Hearken! and observe how healthily—how calmly—I can tell you the whole story.*—Edgar Allan Poe, "The Tell-Tale Heart," 1843

# *Contents*

# *Acknowledgments*

THIS WORK IS THE culmination of many different trajectories of thought and research over the past few years. It is an accomplishment that would not have occurred without the help and support of numerous friends and colleagues. I am especially grateful for the continuous support and insight of Chandra Mukerji, Steve Epstein, Patrick Carroll, Joshua Barker, and Kris Prasad. Over the years I've also received helpful advice and comments from numerous colleagues in science studies. I would like to thank members of the "writing group" and students in the science-studies program at the University of California, San Diego, members of the "monster group" at the University of Toronto, and students and colleagues at Queen's University. I am also thankful for extended conversations on the themes of this book with Malcolm Ashmore, Luigi Bianchi, Harry Collins, Adrian Cussins, Tom Gieryn, Bruce Lewenstein, and Trevor Pinch. I would also like to express my gratitude to the many cold fusion researchers who were kind enough to spend time with me discussing issues in both their work and mine. Our differences of opinion and their ongoing interest in my work provided the fuel that kept this project moving forward. In particular I would like to thank Dieter Britz, Mike Carroll, Eugene Mallove, Hal Fox, Peter Glueck, Michael Schaffer, Massoud Simnad, and Edmund Storms. My final thanks go to my partner, Jill Didur, for sounding out my thoughts, for steering me through, and for making the long journey a worthy one. I am looking forward to wherever the road leads next.

Parts of chapters one and seven have appeared previously in Simon (1999), and a part of chapter 2 appears in Simon (2001). I am grateful to Sage and the Institute of Physics Press for permission to reprint. Figure 7.1 appears with the permission of Martin Rudwick and the University of Chicago Press. Some other quotations appear in the text with the permission of the Cornell Cold Fusion Archive, Rare and Manuscript Collections, Carl A. Kroch Library, Cornell University, Ithaca, N.Y.

# Undead Science

# *Chapter 1* Science Studies and Second Sight

## *Ghosts in the Laboratory*

"Let's walk before we run." This is what RF told me the first time we met to discuss the experiments I would be helping him set up.[1] These were a variation on simple electrolysis experiments familiar to nearly all chemists and chemistry students, like the process used to dissociate water into hydrogen and oxygen gas. You have two electrodes—an anode and a cathode—immersed in a salt solution to make an electrochemical cell. When you pass a current through the cell, molecules of $H_2O$ dissociate into hydrogen and oxygen ions that are drawn to the electrodes. Hydrogen ions move to the cathode, where they recombine into hydrogen gas, and oxygen moves to the anode and bubbles out as oxygen gas. Even for a sociology graduate student such as me, the basic setup was certainly well within my grasp. But we were interested in something much more esoteric than simple electrolysis. We were investigating the possibility of a reaction at the cathode that might produce excess heat and maybe some neutrons.

In 1989 there had been a big scientific controversy about the possibility of such a reaction, which the media and scientists referred to as "cold fusion." RF, a senior materials engineer at a large American research university, knew some of the scientists that had reported measuring excess heat by electrolysis using palladium cathodes in a heavy-water ($D_2O$) electrolyte, and he had managed to do a few experiments with his students, which hadn't led to much. The controversy over cold fusion had basically ended by 1990, but when we met in 1993, RF was still interested in the topic. That is when I started to work with him, doing minor grunt work in the lab off and on for a couple of years while I did research for my dissertation on the history and sociology of the cold fusion controversy. As part of the Ph.D. program, we were required to do internships with scientists to get a better sense of what is sometimes called "laboratory life." I was still interested in the cold fusion case from work I did for my masters thesis, and RF was happy to have some help in the lab, as he couldn't find any other engineering or science students who wanted to work on the topic.

We started with something simple based on Reiko Notoya's 1992 work in Japan, a small, thin piece of nickel foil and a coiled piece of platinum wire immersed in a solution of distilled water and potassium carbonate. We would electrolyze the solution overnight at 1 milliamp and then slowly increase the current to half an amp or an amp for a two-to-three-hour experiment. I read the temperature of the solution off a couple of thermistors every ten minutes, but we also had an old circular chart recorder tracking temperature changes at a shorter interval. It wasn't the prettiest setup. We were using small Rubbermaid containers and bits of platinum wire, and it was difficult to get the geometry of the cathode and anode right. We were sort of shooting in the dark, but we had the reports of Notoya (1993) and also Mills (1991) to work with. We didn't expect to see anything, but RF wanted to get some experience with electrochemistry and think about possible cell configurations and other parameters without wasting money on more expensive palladium cathodes and the more dangerous lithium deuteroxide electrolyte.[2] We ran quite a few experiments using light water and nickel, and the temperature plots looked just as they should; at each interval of applied current, the electrolyte temperature would increase and then stabilize, always within predicted values. On occasion we did notice the temperature spiking higher than it should, and while I was certainly excited about it, RF did not think much of it.

Things became a little more interesting when KD joined our group in 1994. KD had access to a portable infrared scanner that he normally used to measure heat diffusion in the walls of nuclear reactors. KD's scanner and expertise brought new possibilities, and RF redesigned the cell by cutting a hole in the plastic container. He used silicone to attach the nickel foil to the side of the container with the hole so that one side was exposed to the electrolyte and the other side was exposed to the air and the lens of the IR scanner. Now we could measure heat diffusion within the cathode itself. This not only gave us a new source of data, but it was also a novel approach that to our knowledge no else had used. Our apparatus was still fairly crude, but now the data were aesthetically beautiful. We had videotapes of shifting colors in psychedelic patterns that corresponded to changing temperature gradients in the cathode and electrolyte.

We ran more experiments, and everything seemed normal. The IR scanner data confirmed the baselines that we had been obtaining with the thermistors and chart recorder, but all that changed one evening. RF and KD were in the lab together late at night (as bad luck would have it, I was not present); they had done a few runs that week, and RF decided it was time to try a new piece of nickel. When they started the experiment after the normal eight-hour, 1-milliamp charge-up period, they immediately noticed a difference. The temperature was increasing more quickly than it should. They followed our usual protocol of increasing current slowly in a stepwise fashion, but instead of the temperature stabilizing at each new level of current, it just kept climbing all the way from 23 to 39 degrees Celsius (a huge increase relative to our baseline experiments). What was more surprising was that when RF tried lowering the current, it had no effect on the rising temperature. Normally after some delay the temperature should fall with the current, but in this case it didn't. According to RF, if the cell had been insulated, the electrolyte might have very well reached the boiling point. As it was, since it was late at night and RF and KD had both stayed much longer than they had intended, they decided to stop the experiment.

The experience gave our group renewed energy. While the plots of the thermistor data confirmed RF's observation of rising temperature with decreasing current, the IR scanner data seemed even stranger. The scanner also recorded the steep temperature increase, but it was even higher than what was recorded by the thermistors. This indicated that the origin of the heat was the cathode and not the electrolyte, but more interestingly on several occasions the video image erupted in millisecond bursts of color, indicating a temperature shift of up to 15 degrees Celsius and back again in the space of one second. As someone commented upon seeing the video, "It looks like a supernova." These IR "flashes," as RF called them, were completely unexpected, and in his twenty years of experience KD had never seen anything like them before. While the first thought was that the flashes were a result of some electrical glitch in the camera, after checking things out KD became confident that it was not a malfunction or any kind of artifact.

Were the images on the screen caused by cold fusion? Every time I asked, RF said it was unlikely and we would have to repeat the experiment many more times before we could ask anything like that question. But there was no doubt that RF and KD were excited. The anomaly we had observed was significant and warranted further investigation. There were definitely problems though: for one, the thermistor data and the IR scanner data did not correlate (although the fluid dynamics of the electrolyte might be responsible for the difference), but more importantly we couldn't get the same results again.

Every experiment we tried after that was an attempt to reproduce the experience of that night. Occasionally we observed the flashes, but never any temperature shifts of the same magnitude. Soon the other responsibilities of RF and KD got the better of them, and they were able to run fewer and fewer experiments. RF spent some time talking and collaborating with other scientists working on cold fusion, including one scientist who brought his cell to RF's lab, where the IR scanner was used with good results. As for our own experiments, we never worked up the data for publication, and while RF briefly investigated some informal avenues for funding the project on a more serious level, nothing seemed promising. It was frustrating; there on our video screen was this apparition—possibly a false signal, possibly real. For RF, even though he told me that weird things happen all the time in scientific research, the experience confirmed his interest in the ongoing investigation of cold fusion as legitimate science, something that many of his colleagues would not agree with. For me, the experience slowly became less about "seeing" cold fusion and more about opening my eyes to a social world of scientific research that I had barely an inkling of. The efforts of our group did not do much to prove or disprove the existence of cold fusion, but it became evident to me that we were a tiny part of a much larger diffuse effort amongst hundreds of scientists still working to do just that.

## *The Controversy over Cold Fusion*

From the moment it was first announced at a press conference at the University of Utah on March 23, 1989, the reality of cold fusion was a matter of public scientific controversy. The "discoverers" were Martin Fleischmann, a retired eminent electrochemist from the University of Southampton, and Stanley Pons, the chair of the chemistry department at the University of Utah in Salt Lake City. Largely self-funded and working independently, Fleischmann and Pons (along with a graduate student, Marvin Hawkins) claimed to have developed an electrochemical process whereby palladium metal immersed in a heavy-water solution under electrostatic pressure could produce upwards of four times more energy than was required to initiate the process. These energy levels were so high that Fleischmann and Pons could find no explanation to account for them other than that some kind of nuclear reaction was occurring. As evidence, Fleischmann and Pons noted measuring above-background levels of neutrons, tritium, and later helium in their experiments. This was important; neutrons, tritium, and helium are typical products of nuclear reactions, and the particular energy levels claimed are unlikely to have been produced by anything else. When Steven Jones, a physicist at Brigham

Young University in Provo, Utah, announced a seemingly similar discovery featuring neutron measurements, public interest erupted into a kind of "fusion fever" that kept the story on the front pages of newspapers for many weeks. On the face of things it looked as if Fleischmann and Pons had made a major discovery of Nobel Prize proportions. If they were correct, then an important new energy source had been found.

Objections to Fleischmann and Pons's claims surfaced quickly and came mainly from nuclear physicists, who argued that it was astronomically unlikely that a chemical process like electrolysis could produce a nuclear reaction of any kind, let alone a reaction that could lead to commercial energy production. The problem was that the amount of energy observed and the level and kind of nuclear ash (neutrons, tritium, etc.) detected were not reconcilable with any known nuclear process or theory. As a consequence of these objections it was assumed that a) Fleischmann and Pons had discovered something completely novel (most scientists thought this was unlikely), or b) nuclear processes were not the explanation for what they claimed to have observed (this was more likely), or c) Fleischmann and Pons had made crucial mistakes in their experimental procedure, data interpretation, or both. The more Fleischmann and Pons publicly argued for a), the more their critics publicly argued for c). Fleischmann and Pons believed that no other kind of explanation could account for their observations; they admitted that their work was still in its early stages but maintained that there was a real nuclear phenomenon present that needed further investigation. The critics disagreed, and many believed that some basic mistake had been made. Almost everyone agreed that scientists should attempt to reproduce the experimental results in an effort to explain what was happening.

Throughout the spring and summer of 1989, hundreds of groups of scientists around the world attempted to reproduce Fleischmann and Pons's cold fusion experiments.[3] Most failed but some claimed success, and the controversy escalated as scientists began to find themselves being associated with one of two groups, the supporters or the critics of cold fusion. By midsummer, the debate between the two groups became increasingly antagonistic, taking on religious overtones in the media and even at scientific conferences. Scientists who spoke out in support of Fleischmann and Pons or cold fusion became pejoratively known as "true believers," while those who voiced criticisms tended to be classified as "skeptics." From the critics' point of view, belief in cold fusion came to be seen as a matter of blind faith rather than good science. And in the supporters' view, most criticisms seemed to intentionally ignore the experimental facts of the matter in favor of conventional scientific dogma.

By the end of the summer, general scientific opinion tended to favor the skeptics, and it became clear that the burden of proof would lie with the believers, who would be held responsible for providing new convincing evidence for the existence of cold fusion. At the same time, however, institutional support and funding were being withdrawn from cold fusion research projects, and by the end of the year, experimental work in cold fusion was in steep decline, leaving perhaps one hundred or so researchers working worldwide. Under scathing criticism, including charges of fraud and methodological impropriety, and without evidence of a reliably reproducible experiment or a working device like a water heater, the case for cold fusion floundered.

In the end, the opinion of most scientists is that Fleischmann and Pons and the other believers were mistaken in the interpretation of their experimental data. The reported signatures of excess energy and nuclear ash were most likely artifacts, and, although this could not be decisively proved, the lack of detection by well-respected experts suggested that some kind of "experimental artifact" hypothesis was the best explanation. Short of that, it was suggested that the cold fusion effects were the products of wishful thinking or a kind of "pathological science." A few of the more vocal critics went even farther and accused some cold fusion researchers of not only ignoring their own negative evidence, but of willfully manipulating data in an attempt to deceive others.

## *The Empirical Problem*

Scientists, journalists, and scholars agree that the controversy was effectively over by the spring of 1990. Fleischmann and Pons's cold fusion hypothesis had been refuted, and the assumption was that the number of scientists working on cold fusion would continue to decline as people gradually lost interest. And that is generally what happened. Federal funding for cold fusion experiments in the United States was withdrawn, papers were no longer accepted for publication in many major journals, patent applications were denied, and scientists returned to other projects. One scientist, David Goodstein, summarized the affair as follows: "Cold Fusion had been given its chance, a suspension of disbelief no matter how unlikely it seemed, and it had failed to prove itself. Cold Fusion was dead in the eyes of respectable science" (Goodstein 1994, 1). A chemist whom I interviewed explained the situation to me this way:

> The problem with cold fusion as a legitimate research goal is that so much of the phenomenological basis for the effect has been shown to be either mistaken, possibly fraudulent, or simply not reproducible in

> the hands of established authorities, that little or nothing remains of the original hypothesis. We remember the excitement of April 1989, the Utah press conference, the "discovery" of tritium, etc., but stripped of these events outside of the normal peer review process, little or nothing remains of the phenomenon. The Utah press conference forced leading theoreticians and experimentalists to reconsider the idea, and no reasonable expectation of this phenomenon resulted. Thus the exclusion of CF from normal scientific discourse . . . seems reasonable—it being an improbable idea with no accepted evidence. Compound that with the allegations of fraud, the existence of unusual "investment opportunities" and other things at the margins of usual science, and CF is clearly a suspect field.[4]

Based on the trajectory of the controversy in 1989 and 1990 and comments from scientists like those above, it certainly would appear that the debate has ended and cold fusion is dead as a legitimate scientific idea, and yet at the same time it could be argued that experimental work on cold fusion has actually not declined significantly.

In fact, the number of cold fusion researchers has remained fairly constant since 1990, and this is coupled with a steady increase in experimental evidence supporting Fleischmann and Pons's cold fusion hypothesis. In addition, cold fusion research has taken on many of the trappings of what we might expect from normal science: yearly conferences, specialized journals, experimental collaboration and proliferation, the development of theoretical frameworks, and an increasing collective confidence in the phenomenon. As the abstract from one scientific literature review of the field in 1996 reads, "Over 170 studies reporting evidence for the 'cold fusion' effect are evaluated. New work has answered criticisms by eliminating many of the suggested errors. Evidence for large and reproducible energy generation as well as various nuclear reactions, in addition to fusion, from a variety of environments and methods is accumulating. The field can no longer be dismissed by invoking obvious error or prosaic explanations" (Storms 1996a, 185). If scientists were indeed still actively doing research on cold fusion and addressing the problems encountered in 1989, then, contrary to popular scientific opinion, one would suspect that controversy must not be over. Cold fusion must still be alive.

The problem, in the first instance, is an empirical one that should be of interest to both scientists and students of science alike. How do we explain contradictory evidence of the life and death of a scientific phenomenon? Despite the fact that cold fusion papers are often rejected for presentation at professional scientific conferences, since 1990 around two hundred researchers working in the field have gathered yearly for the International Conference on Cold Fusion (ICCF). There have also been numerous workshops and seminars, and there are a number of magazines and newsletters that report regularly on developments in the field. Funding is difficult to come by, but there are important exceptions. From 1993 to 1998, for instance, the Japanese government sponsored the New Hydrogen Energy Research Project, a $30 million program directed toward the commercial development of cold fusion technology. There are also numerous smaller research projects at universities and private companies in the United States, France, Russia, and Italy, all claiming to have experimental evidence in support of cold fusion. Perhaps the most fundamental criticism of cold fusion is the apparent lack of a reliably reproducible experiment, but within the last two years at least three companies have announced the development of commercializable heating devices employing cold fusion. If cold fusion is dead, then how shall we account for these observations?

## The Analytical Problem

In the second instance, the problem is an analytical one having to do with how we should approach the study of scientific controversies and their resolution. The development of cold fusion research in the wake of the controversy of 1989 presents scholars in science studies with something of a dilemma, which we might call the social-scientific (rather than the natural-scientific) problem of post-closure cold fusion. This is to say that the problem is a methodological one for sociologists and historians of science (along with the anthropologists, philosophers, economists, and literary scholars who work in the interdisciplinary field of science studies). Most scientists would not experience a problem in this sense. As one physicist commented on the recent public announcement that a cold fusion water heater had been developed, "I didn't buy it then, and I don't buy it now."[5] From this physicist's perspective, cold fusion is dead, and those who continue to do research on the subject are part of an incompetent or sadly misguided minority of scientists who may or may not discover their error. By the same token, scientists who are still working on cold fusion also do not experience this social-scientific problem since they believe that cold fusion is real (if only in the sense of not being an artifact). Cold fusion researchers believe that many skeptics are simply blind conformists who are unable to accept the accumulating evidence before their eyes.

The problem then, to the extent that there is one, is purely a science-studies problem. How do we account for the perspectives of the supporters and the critics and everyone in between, and still make sense of the outcome of the controversy? In other cases, this kind of problem does not arise. Historical studies of controversy are usually informed by hindsight. We know when a study starts what the outcome was, and the task becomes a matter of accounting for how that outcome was achieved. Even in contemporary controversies, there is usually a high degree of convergence or agreement amongst participants in the end. So much so, that to the extent that such a convergence is not observable, our inclination is to assume that the controversy has not really ended.

The assumption here is not that scientists ultimately always agree with one another. We recognize that varied and even contradictory beliefs about the world are not uncommon amongst scientists, but at the same time we have come to understand controversies to be processes leading to resolution. The diverse interests, beliefs, and practices of various scientific actors eventually give way or converge to a set (or at least multiple sets) of common collective interests, beliefs, and practices—a shared "form of life" (Collins 1985; Shapin and Schaffer 1985) or "social world" (Strauss 1978; Clarke 1991). Controversies, in this sense, become an important social means through which epistemic order and scientific truth are established and sustained (Rudwick 1985, 5; Collins 1981c).

In the sociology of science, the most prominent conceptualization of this comes from Thomas Kuhn's notion of a paradigm in which social and epistemic instability or divergence lead eventually to the stable practices and beliefs of normal science (Kuhn 1970, 17). Beyond Kuhn, this idea is an aspect of many approaches in contemporary science studies, where scientific controversies are thought to be productive of both stable social worlds and knowledge. Post-Kuhnian constructivist sociologists of science describe this process in terms of the coordination of action (Barnes 1982), the black-boxing of facts (Latour 1987), or the putting of ships into bottles (Collins 1975). All these metaphors refer to the simultaneous construction of knowledge and social order in terms of the collective convergence of practice and belief, and the elision of difference.[6]

For science-studies scholars, the trick is in deciding when sufficient agreement has been reached to constitute the end of a controversy, or closure. The idea is to account for when and how controversies end, and what their endings consist of; who are the winners, who are the losers, and what does the new landscape of scientific knowledge look like? It is in this process that constructivist science studies continue to be guilty of some interpretive whiggishness. Typically, attributions of closure by science-studies scholars are meant to follow the performance of closure by scientists. That is, controversies end once the scientists end them. Closure is recognized in this sense by the stabilization and routinization of practice, and by the near absence of conflict amongst scientists.

The consequence of this for science studies is a dualistic view of the production of scientific knowledge. This view is neatly expressed in Bruno Latour's metaphor of the Janus face: "If you take two pictures, one of the black boxes and the other of the open controversies, they are utterly different. They are as different as the two sides, one lively, and the other severe, of a two-faced Janus. 'Science in the making' on the right side, 'all made science' or 'ready made science' on the other" (Latour 1987, 9). This metaphor informs part of the background assumption of Latour's "first rule of method" for studying technoscience: "We will enter facts and machines while they are in the making; we will carry with us no preconceptions of what constitutes knowledge; we will watch the closure of the black boxes and be careful to distinguish between two contradictory explanations of this closure, one uttered when it is finished, the other while it is being attempted" (Latour 1987, 14–15). Latour's rule of method reinforces the view that solutions to the problem of knowledge are solutions to the problem of social order (Shapin and Schaffer 1985, 332) and that in the absence of social order there is an absence of knowledge (expressed as a collective inability to know); social stability produces certainty.

In the study of scientific controversies, order and knowledge are the twin products of processes of closure. This makes the understanding of closure a core problem in the sociology of scientific knowledge (Collins 1981d; Galison 1987). In order to understand closure and trace its accomplishment, science studies has adopted this Janus-faced or "stereophonic" (Latour 1987, 100) view of knowledge making. But the case of cold fusion, unlike most others, threatens to give us whiplash as we throw our heads in opposite directions trying to decide if cold fusion is alive and the case is open or dead and the case is closed. The empirical situation seems to suggest that neither position is adequate. The task of this book is to try to find a satisfactory position and resolve the anomaly of cold fusion, not for scientists (who can handle this problem on their own), but for science studies.

## *Methodologies and Metaphysical Presuppositions*

Starting from Latour's first rule of method, we have only two options for resolving the social-scientific problem of cold fusion. The first option is to find that the rumors of the death of cold fusion have been greatly exaggerated and the controversy continues (albeit in a different form). The second option is to argue that the vital signs of cold fusion are weak at best, and the "experts" are confident that it will pass away for good any day now. We could hope for some kind of remission, but what would be the point? As science-studies analysts that is not our job. Even the experts recognize that there is always a possibility of remission (after all, the history of science is replete with surprising turns of events), but they suggest it is better to prepare for the inevitable, and who are we to disagree? As Latour counsels,

> When talking about a cold part of technoscience we should shift our method like the scientists themselves who, from hard-core relativists, have turned into dyed-in-the-wool realists. Nature is now taken as the cause of accurate descriptions of herself. We cannot be more relativist than scientists about these parts and keep on denying evidence where no one else does. Why? Because the cost of dispute is too high for an average citizen, even if he or she is a historian or sociologist of science. If there is no controversy among scientists as to the status of facts, then it is useless to go on talking about interpretation, representation, a biased or distorted world-view, weak and fragile pictures of the world, unfaithful spokesmen. Nature talks straight, facts are facts. Full stop. There is nothing to add and nothing to subtract. (Latour 1987, 100)

From this, our method seems clear. We should talk to the experts, watch how they interpret the vital signs, and ask questions. As long as they bicker and disagree, we will stand watch. When death is pronounced, we should not be relativists and contest it. On the one hand, many experts say that cold fusion is dead, but on the other hand we can always find scientists who will disagree. Latour has left us with the knotty problem of figuring out how many dissenting experts it takes to keep a controversy alive. This is a problem he solves in part by adding nonhuman actors to the mix, which may turn the tide in "trials of strength" that punctuate the agonistic dynamics of controversies (Latour 1983, 1990). On Latourian battlefields, as with much of the sociology of scientific knowledge, there are always winners and losers.

But what if death is not the end of the story? In science studies we are typically skeptics and agnostics (Latour 1981), and while we may be radical enough to attribute "variable ontologies" to phenomena in controversies as they come into being, live, and die, we have nothing to say about what, if anything, lies beyond.[7] Perhaps open controversies and closed controversies are not all there is? Cold fusion is dead. We can prove this, and we will not contest it for the reasons Latour mentions above. But I will suggest that cold fusion is also very much alive. That which is both dead and alive is, to put it bluntly, undead. This implies a new analytic category for social studies of science: cold fusion is neither dead nor alive; it is undead, like a ghost, a phantom, or a specter. If we admit the possibility of ghosts, we can resolve the problem of post-closure cold fusion while still recognizing the force of closures and the effects of "losers" without becoming relativists. The argument that follows in this book is intended as a justification and beginning of just such an eschatological investigation, something like a hauntology for the technoscientific afterlife.[8]

My main point, which is only a starting point, is that being undead is neither an intermediate nor a nonexistent (or imaginary) state of being. Ghosts occupy the boundaries or border zones between the living and the dead. The living are the real things and legitimate people of our social worlds, while the dead are the failed remnants (people and things) of old controversies, which no longer exist and cannot act. The undead, however, do exist and can act, but their actions are mediated and constrained by their relations with the living. This is a consistent trope in the literature of gothic horror; ghosts are frequently tied to houses or people, vampires depend on the blood of the living, and so on (Botting 1996). Thus, cold fusion exists, but its existence is circumscribed in the first instance by its death in 1989–1990, and in the second instance by its relations with legitimate science after 1990.

At the same time, the post-closure existence of cold fusion has repercussions on the production of legitimate scientific knowledge. As ghosts, cold fusion researchers may haunt the laboratories of the normal institutions where they work, their practice affecting the practice of legitimate research as ideas, people, and materials continue to circulate. Ghosts are ephemeral and invisible, they come and go, but their actions may have important social, material, and epistemological effects on the worlds in which "we" live (think of poltergeists and haunted houses). As a consequence, being undead should not entail being forgotten, being dismissed, or being unimportant, even in an explanation of the production of scientific knowledge. The case of cold fusion is our entrée to the worlds of undead science, and this book is meant to help make analytic space for all manner of technoscientific ghosts and monsters in the sociology of scientific knowledge.[9]

My intention is, for the most part, to leave the argument for a hauntological perspective in science studies until the final chapter, and even then the project of developing a full-blown hauntology is left for a future work. I see my main task as establishing an empirically informed conceptual groundwork for being able to see ghosts in the first place. The argument for abandoning our stereophonic life-or-death metaphysics has yet to be made, and in order to do so we will need to work with analytical tools we already possess.

## Methodological Symmetry and Postclosure

In *Knowledge and Social Imagery*, David Bloor (1991) sets out, in part, to open the door of traditional epistemology for sociologists and historians, and to make a sociology of scientific knowledge (SSK) possible. A central element of Bloor's argument is the epistemological importance of the deconstructive moment implied by taking on a position of methodological symmetry when studying science. According to Bloor, the sociology of scientific knowledge would not be just a sociology of scientists; it would explain the cognitive content of science (scientific knowledge) in sociological terms. It would do this by being "impartial with respect to truth or falsity, rationality or irrationality, success or failure." And "it would be symmetrical in its style of explanation. The same types of cause would explain, say, true and false beliefs" (Bloor 1991, 7).

In the absence of this "style of explanation," accounts of science become asymmetrical. That is, true beliefs tend to be explained in terms of rational or natural causes, and false beliefs are explained in terms of social or psychological causes. As a consequence, asymmetrical explanation limits the possibility of a sociology of knowledge to a sociology of error. In the case of cold fusion, for instance, we may be called upon to account for Fleischmann and Pons's mistaken beliefs: were they influenced by political or economic forces beyond their control? Or perhaps the reasons for their failure can be found in their individual personalities. A position of methodological symmetry grants the possibility of this kind of explanation, but only if political and economic forces or individual personalities can explain true beliefs as well.

It is important to note that methodological symmetry does not necessarily imply epistemological relativism (Collins 1981b). Rather, it is a perspective that opens up the analytical terrain to an understanding of scientific knowledge as the product of coordinated social-material action. One of the most common ways of doing this is by looking at the social dynamics of scientific controversies. Without importing presumptions about who is right and who is wrong, how do controversies get resolved? How is agreement achieved on what constitutes the "fact" of the matter? The regulative assumption here is simply that any historical outcome of a controversy could have been otherwise; there is no a priori truth of the matter for actors to agree with, and the truth of the matter becomes a consequence of their agreement. If actors agree or arrange themselves differently, then the truth of the matter is different. But there is ultimately still a truth of the matter; human beings as social beings cannot believe just anything they like.

So the notion that controversies could have ended differently must be understood simply as a regulative assumption for the sociology of scientific knowledge. It is an "in principle" argument justifying methodological symmetry, not a proclamation of epistemological anarchy. Critics of science studies have consistently misunderstood this point. To say it could have been otherwise is not to say that it is otherwise, and to understand science sociologically is to understand why this is so. We only say it could have been otherwise knowing full well from our point of view that is wasn't otherwise, that it isn't otherwise, and that we as members of our society have a very good sense of who were the winners and who were the losers in controversies. We know what is true and what is false, and we live with and in the consequences of these controversies. Indeed, our sense of all this is as good as, or no different from, that of the whiggish historians and positivist sociologists who are opposed to the sociology of scientific knowledge in the first place.

It is for this reason that the principle of methodological symmetry does not apply beyond the point of closure, the point where "could be otherwise" becomes "is not otherwise." This becomes the problem in terms of accounting for post-closure cold fusion. Methodological symmetry allows us to analyze the cold fusion controversy on its way to closure; it provides a means of sociologically accounting only for closures, for the production of "truth" and "falsehood." Any attempt to symmetrically analyze beliefs after closure would threaten the basis of the sociology of scientific knowledge, as it is predicated on the assumption that closure occurs. Thus in his brief discussion of the outcome of the controversy over the existence of N-rays, Bloor writes, "Sociologists would be walking into a trap if they accumulated cases like Blondlot's and made them the center of their vision of science. They would be underestimating the reliability and repeatability of its empirical base; it would be to remember only the beginning of the Blondlot story and to forget how and why it ended. Sociologists would be putting themselves where their critics would, no doubt, like to see them—lurking amongst the discarded refuse in science's back yard" (Bloor 1991, 30). The outcome of the controversy over N-rays went against Blondlot, as it did for Fleischmann and Pons. Bloor and Latour would agree on this point.[10] To pursue a symmetrical post-closure study of N-rays or cold fusion would be the equivalent of proclaiming that fairies are real. It would make sociology into a laughingstock. The idea of methodological symmetry is not to prolong controversies; it is to understand their closure. When the controversy ends, we must leave methodological symmetry at the door.

## *The Hauntological Method and Second Sight*

The view I propose accepts this, but advocates what I will call a methodological double take on cold fusion as a refinement or diffraction of Latour's stereophonic vision. My pocket dictionary usefully defines *double take* as "a second look given to a person, event, etc., whose significance had not been completely grasped at first." Let the first glance be a symmetrical glance that treats the observation of continued cold fusion research after 1990 as being just like any other knowledge-making culture. In spite of what experts like Bloor say, the production of what counts as science, and the production of knowledge about cold fusion after closure, must be treated as comparable. The second glance, however, will be asymmetrical. In spite of the continuing dissent of cold fusion researchers, we will unflinchingly treat cold fusion as bad or pathological science, just as Bloor does in the case of N-rays. The idea here is that this double looking provides a method for seeing beyond closures that resists the need for a diagnosis from the experts without falling prey to epistemological relativism.

My position is thus different from that of Malcolm Ashmore (1993), who takes issue with Bloor on Blondlot in another way. In his own study of the N-ray case, Ashmore approaches the failure of Blondlot symmetrically by analyzing the discourse of Blondlot's chief critic, R. N. Wood, in an attempt to deconstruct Wood's credibility and the basis for the historical rejection of the N-ray claims. Ashmore's analysis effectively unsettles a historical closure, and in a sense he is doing the work that Blondlot and his supporters might have attempted to do at the time. It is an instructive exercise, but my project is substantially different. The force of my argument depends on my being able to demonstrate that closure has occurred in the case of cold fusion. Not only is my ability to unsettle such a closure limited (how many scientific experts would believe any claims *I* could make that cold fusion is real?); it obscures the view I hope to describe. Without a notion of closure, I will argue, we will not be able to make sense of how cold fusion research manages to survive. Without a notion of closure, cold fusion is simply alive, and much of what has occurred in the case will not make any sense.

The methodological perspective I propose is not like the Latourian Janus face that looks in opposite directions, toward "science in the making" on the one hand and toward "ready made science" on the other. Latour asks us to shift between being relativists and being realists as scientists do and so to avoid being either. The perspective I propose instead demands a double take that looks twice in the same direction, modifying each view in light of the other. In the symmetrical glance, cold fusion is alive and we are relativists. In the asymmetrical glance, it is dead and we are realists. For Latour we must flip-flop between one state and the other; I suggest we must see both states at the same time. Those who see ghosts are often said to be gifted (or cursed) with the ability of "second sight." I will not lay claim to any such ability, but the point is well taken. Seeing ghosts is a function of one's conceptual apparatus. The double take is a heuristic that does not so much expose reality as it allows us to entertain the possibility that something unusual might be going on.

## *The Sociology of Rejected Science*

This fixation on ghosts and the concept of undead science is not meant to be a flight of fancy. I have been prompted on the one hand by the constant talk of scientific life and death on the part of the scientists I have talked to, and on the other hand by sociologies of knowledge that seek to make sense of epistemological plurality and difference without falling back on epistemological relativism. There are other options however. Without making reference to ghosts, we can find another source for thinking about the analytical problem of cold fusion in the literature on the sociology of "rejected" sciences. Interest in such studies blossomed in the late 1970s and early 1980s as the practices of acupuncture, creation science, parapsychology, ufology, mesmerism, and phrenology amongst others were reconsidered in light of methodological symmetry and the sociology of scientific knowledge.[11]

The main point of many of these studies is that the rejected and marginalized claims of parapsychologists, for instance, are not a priori irrational or at all unscientific, and indeed the classification of these claims as "bad" or "pseudo" science is part of the social process that leads to their rejection. Yet in addition to documenting the social-historical dynamics that lead to closure and rejection, these studies laid the groundwork for making sense of contradictory perspectives within a notion of science defined in terms of its social dimensions (rather than its imputed rational properties). One way of doing this is to draw on theories of deviance in sociology to describe the relation between normal or orthodox science and deviant science. R.G.A. Dolby, for instance, writes that

> orthodox science is that which commands the approval of all the leading scientific experts of the time. It includes all the historically successful sciences. Deviant science is that which is rejected by the orthodox scientific experts, and which they may label "pseudo-science"; however, it has its own body of supporters, who claim it to be a science. The terms "orthodox science" and "deviant science" are not purely descriptive, for they involve a social evaluation, as is clear when there is a disagreement about who the experts really are. But in most cases of modern *natural* science, the distinction between orthodox and deviant science is unproblematic. (Dolby 1979, 11)

While the concept of deviance has gone out of favor because of its evaluative overtones, Dolby's classification points to the ways that alternative and contradictory scientific beliefs may be sustained indefinitely by social groups as opposed to just a few stubborn individuals. These deviant currents within science may have a limited infrastructure (Allison 1979) and may become the source of renewed attention by the orthodoxy (Webster 1979; Le Grand 1988), but in all cases their experience is marked by the evaluation or dismissal of a ruling scientific orthodoxy that is recognizable to most observers. What the category of deviant science did was to provide an analytical route to making sense of failed scientific claims that did not just disappear and yet seemed clearly distinct from normal institutionalized science.

As the theories and methods in science studies developed, the interest in the specific dynamics and organization of rejected or deviant sciences declined. There are perhaps a number of reasons for this. One has to do with Bloor's concern about sociologists "lurking amongst the discarded refuse in science's back yard." The case for the sociology of scientific knowledge, still trying to gain legitimacy in the academic arena, would not be made on studies of N-rays or paranormal spoon bending but rather on the harder cases of solar neutrinos (Pinch 1986a) and quarks (Pickering 1984). Another reason has to do with the increasing focus on microsociological studies of scientific discourse and practice (Latour and Woolgar 1979; Knorr-Cetina 1981). These "laboratory life" studies produced a picture of science as situated, pluralistic, and heterogeneous, eliminating the need to define distinct structural categories for science. The notion of deviant science became less of a kind of social organization and more of a rhetorical resource used by actors in local demarcation battles. Thus the focus shifted from the characteristics of rejected sciences in themselves to the process of rejection, a process that has become known in the literature as boundary work (Gieryn 1983).

The argument of this book is heir to the microsociological focus of recent science studies, but my debt to the early work on rejected science will be obvious. I have abandoned the categories of rejected, deviant, and marginal science in part because, in light of recent science studies, the boundaries that analytically mark the distinction between orthodoxies and alternatives in the sciences no longer seem solid. The focus on rejection as boundary work suggests that both orthodoxy and deviance in science must be made and remade on a continual basis and that because of this the line that separates the two becomes quite porous. I might suggest that undead science is a category of practice situated somewhere between orthodoxy and deviance, but that is not quite right. The metaphor of the ghost is meant to capture the distinction implied by the idea of deviant science while pointing to the ways in which the "otherness" of deviant science is present in the normal life of orthodox science. Deviant science in this sense does not exist in a world apart from orthodox science, but rather is mixed up with it, an integral though invisible element of the modern technoscientific world in which we all live.

## *Why Study Cold Fusion?*

This book is about the scientific controversy over the existence of cold fusion, its closure, and its aftermath. In this respect, the book resembles and engages with a number of other case studies of scientific controversy in the sociology of scientific knowledge (Collins 1975, 1981c; Collins and Pinch 1982; Pinch 1985; Rudwick 1985; Shapin and Schaffer 1985; Latour 1988a; Pickering 1980). My approach to the study of this controversy generally adheres to common principles of method outlined by Harry Collins (1981b, 1983, 1985) in the empirical program of relativism (EPOR) and actor-network theory as articulated by Latour (1983, 1987), Callon (1980, 1986), and Law (1986).[12] While there remain crucial points of disagreement between the EPOR approach and actor-network theory over questions of the distribution of agency amongst human and nonhuman actors (Collins and Yearley 1992; Callon and Latour 1992), this case study assumes a basic continuity in both approaches to the study of controversy. Both EPOR and actor-network theory treat knowledge as the product of local struggles that result in an eventual convergence in the constitutive practices of actors (whoever or whatever those actors happen to be).

To this extent, this case study serves both as further "replication" of other similar kinds of studies (Collins 1981a, 4) and as an opportunity to investigate and evaluate extant analytical tools for studying science in light of a new case. My argument is not intended as a break with or criticism of existing approaches so much as an extension or broadening of the categories of analysis for the study of controversies. This may seem overly pedantic, but my intention is to throw new light on old controversy studies by investigating the limits of routinely deployed concepts in science studies, such as the very terms "controversy" and "closure."

For social scientists, philosophers, and historians already predisposed to relativistic and constructivist approaches to the study of science, case studies of controversy have become standard pedagogical tools. They are used as exemplars of the basic tenets of the sociology of scientific knowledge and technology, as well as models for training in the methodology of social-science research.[13] Controversy studies have become so standardized in this sense that some scholars have even begun to question their use for advancing research in the field (Hacking 1993). As a pedagogical tool, the study of the controversy over cold fusion is indeed exemplary. The mass-media exposure of the case in the spring and summer of 1989 meant that an overwhelmingly large number of people became familiar with the technical details, the arguments, and the evidence upon which the controversy hinged. This provides science-studies scholars with a perfect empirical touchstone for supporting theoretical claims. Although the controversy was short-lived, for a time the outcome was completely unpredictable as highly reputable scientists from around the world argued for and against the reality of cold fusion through newspapers, magazines, and scholarly journals.

For students of controversy the case of cold fusion is a gold mine. Both the magnitude of the controversy in terms of the number of actors involved and the ability to access materials produced by those actors have provided a wealth of data on scientific controversy as a social phenomenon. At one point, the controversy was deemed important enough for the National Science Foundation to support the development of an archive of materials related to the case at Cornell University (Lewenstein 1991). In addition, in the several years since the main controversy ended, a small publishing industry has developed around the topic. While this is most evident in the various media of popular science, a significant number of scholarly works have been produced as well.[14]

While popular work on cold fusion certainly serves as further fuel for analysis, it can perhaps be argued that previous scholarly attention has all but exhausted the topic for science studies. I will argue that the opposite is the case because previous scholars have failed to note the ways in which the work on cold fusion has continued after the end of the controversy. The limit of analysis in most other controversy studies is the end of the controversy; this book is an attempt to look beyond that limit. The case of cold fusion has been studied insufficiently in part because its significance is often deemed to be limited to that of a pedagogical tool, an empirical occasion that simply reinforces what science-studies scholars already know (Lewenstein 1992b; Collins and Pinch 1993, chapter 3). This perspective overlooks what amounts to an important opportunity to build upon the constructivist paradigm in the sociology of knowledge without having to reiterate first principles.

## *Follow the Object*

The method I have employed to study this case can be summed up in the single imperative often iterated by Bruno Latour to "follow the object." For Latour, cold fusion might constitute a kind of "quasi-object" whose ontological status is in a state of flux (Latour 1993, 1996). What the object "cold fusion" is at any given point is a product of the network of human and nonhuman agents that are associated with it. Thus, to follow the object is to trace its network of associations with no a priori conception of which associations are more or less significant than others. In practical terms, I have interpreted my task as tracing the movement and deployment of the signifier "cold fusion" by accounting for its appearance, nonappearance, and transformation in specific contexts.

To accomplish this, I have made use of a wide variety of primary sources, ranging from interview and participant observation data to newspaper articles, technical papers, conference talks, newsletters, private and public e-mail, letters, government documents, patent applications, advertising, and even Hollywood films. Much of this material I collected in the course of talking with scientists, but I also made use of material housed in the Cornell Cold Fusion Archive.[15] I have also made extensive use of the annotated bibliography of technical papers on cold fusion compiled by Dieter Britz (a chemist at Aarhaus University, Denmark), a full set of the newsletter *Fusion Facts* supplied to me by Hal Fox (the director of the Fusion Information Center in Salt Lake City, Utah), and the ongoing discussions on two Internet listservs related to cold fusion.

The background research for this book was accomplished in conjunction with a participant observation study of a small research group working on cold fusion at the large research university mentioned at the beginning of the chapter. I served as an occasional research assistant, helping to design and carry out cold fusion experiments as well as conduct literature searches and analyze experimental data. In addition, I attended the fourth International Conference on Cold Fusion held in Maui, Hawaii, in December 1994, as well as the second Cold Fusion Workshop held in Asti, Italy, in December 1997.

Finally, additional data are provided from a series of depth and incidental interviews with various participants in the cold fusion controversy. I conducted depth interviews from one to two hours in length with twenty CF researchers in the United States, Canada, France, Italy, and India from 1994 to 1997, and I engaged numerous others in shorter conversations and e-mail exchanges over this period. Additional material concerning the early history of the controversy was obtained from interviews with scientists conducted by Tom Gieryn and Bruce Lewenstein (these are housed in the Cornell Cold Fusion Archive). A number of respondents wished to remain anonymous in this study, so I have elected to maintain the confidentiality of all respondents. Where I have used publicly available data, identities will be supplied for the purpose of developing the historical record, but in the case of interview data only the professional identity and the date of the interview will be noted (e.g., interview with a chemist, March 3, 1996).

## *Summary of Chapters*

The first part of the book discusses the history of the cold fusion controversy and its relevance to issues in the sociology of scientific knowledge. Chapter 2 explores the early part of the controversy by focusing on the configuration of cold fusion through the media and the technical literature, and the response of scientists. The main argument of the chapter is that neither the phenomenon itself nor the group of scientists who became involved in its validation are given at the start. Rather, the phenomenon, the methods for its validation, and who would count as an expert in this case are configured through various representations. These representations give researchers a sense of what they are supposed to be looking for and how they are supposed to evaluate what they find.

The third chapter traces technical aspects of the controversy and the actions of a core set of scientists engaged in the experimental verification of the effects associated with cold fusion. The various arguments for and against the existence of cold fusion are discussed, and the failure to reach consensus offers further support for the argument that scientific theories are underdetermined by evidence. The resolution of the controversy is brought about not by the rational application of reason or method, but by a convergence of belief coupled with a reduction of "interpretive charity," processes of discrediting and delegitimation, and the withdrawal of material resources for research. The argument of this chapter is that closure in the case of cold fusion has to do less with the agreement of actors than with the role of legitimacy and resources in allowing the participants to act. Our understanding of closure as the alignment of belief or practice should be qualified to take into account the social and material conditions that make continued collective dissent either possible or impossible.

Chapter 4 analyzes the post-closure reconfiguration of cold fusion as a case of what the Nobel laureate Irving Langmuir called pathological science, or "the science of things that aren't so." Cold fusion was not simply disproved, but actually proved to be an experimental artifact. As long as cold fusion was deployed as an artifact, anyone who continued to pursue cold fusion research could be less problematically labeled as incompetent. In the wake of the controversy, the increasing entrenchment of cold fusion as pathological science made continuing resistance to the closure of the controversy more difficult while at the same time reinforcing conventional concepts and practices associated with the study of nuclear processes. This chapter analyzes the discourse of pathological science as a form of boundary work and looks at its epistemological effects in terms of the destruction and reconstruction of the phenomenon.

The fifth, sixth, and seventh chapters examine the organization, beliefs, and practices of cold fusion researchers working after the closure of the controversy of 1989. The analysis in these chapters focuses on how CF researchers as a group manage to sustain and develop a research program in the face of opposition from mainstream scientists. The curious position of CF researchers is reflected in a tension between the pursuit of a coherent experimental program designed to make the anomalous effects more reproducible and the extension of the properties of the phenomenon to include even more heterodox claims. Particular attention is directed to issues of shared experimental practice, resource mobilization, communication, and the formation of collective identity.

The final chapter picks up strains of the theoretical and methodological argument from chapter one. The case of cold fusion unsettles conventional understandings of the dynamics of controversies in science studies in terms of a convergence of belief and practice. The idea of closure as convergence is discussed and criticized in light of the case of cold fusion. The continued survival of CF research suggests that the production of scientific knowledge goes on outside the normally convergent networks of legitimate science, and that in order to have a more complete sociology of knowledge, analysts must take these underground networks of undead science into account.

# *Chapter 2* The Birth of Cold Fusion

## *Spectacular Science*

Just after midnight on March 24, 1989, the oil tanker *Exxon-Valdez* ran aground near the Alaskan coastline in Prince William Sound, spilling over 10 million gallons of crude oil into the water. The resulting ecological, political, and financial disaster served as a wake-up call alerting the North American public both to our dependence on fossil fuels for energy and the crises that such dependencies can precipitate. The story of the spill was spectacular, not only for its magnitude (it was the largest single oil spill in history) but also for its poignancy. Images of oil-soaked birds deployed by environmental activists and the media highlighted the need to develop a "greener," more responsible approach to the production and distribution of energy.

Coincidentally, just hours before the *Valdez* ran aground, major newspapers around the world were reporting a startling new scientific discovery that might eliminate the possibility of oil spills. As noted in chapter 1, two chemists working at the University of Utah had supposedly discovered a new form of clean, inexpensive energy; the energy was produced in a strange chemical-nuclear process that would become known as "cold fusion." Like the process of nuclear fusion, to which it was apparently related, if cold fusion could be commercialized, it would likely transform the cultural and economic infrastructures of modern societies in profound ways. At first glance cold fusion appeared to be the key to some kind of utopian world dreamt up in a science-fiction novel.[1]

The spring of 1989 has since been etched into the collective memory of scientists and futurists alike. In the media and in the public imagination, dystopic narratives of ecological catastrophe were suddenly juxtaposed with millenarian speculations about the impending energy revolution. This was a modern American tale of the triumph of bench-top science and individual invention (Toumey 1996). While environmentalists were arguing against the dangers of technocracy, the patient, unassuming scientists were becoming heroes. Science would triumph as new technologies brought progress and hope to society in the form of an energy revolution.

This chapter examines the first representations of cold fusion and their reception by the scientific community. Two kinds of representations are of particular interest, the "public" representations of press conferences and newspaper reports and the more esoteric technical representations of scientific journal articles. Both kinds of representations importantly configure the phenomenon of cold fusion by not only giving audiences a sense of what cold fusion is, but also circumscribing who is qualified (and who would qualify themselves) to determine if the phenomenon is real. Consequently, the initial representations of cold fusion are an important part of what makes the claims for its existence controversial.

## *Core Sets and Scientific Controversy*

Generally speaking, scientific controversies occurring at the frontiers of research usually involve only a very small subset of the general population of scientists. This subset is sometimes referred to in science studies as a *core set*, and it consists of those scientists "who are actively involved in experimentation or observation, or making contributions to the theory of the phenomenon, or the experiment, such that they have an effect on the outcome of the controversy" (Collins 1981b, 8). The core set is typically composed of informed specialists doing esoteric technical work who, directly or indirectly, engage with one another for the express purpose of resolving the scientific dispute.[2]

For obvious reasons, science-studies scholars have tended to view core sets as the basic locus of knowledge making in scientific controversies. Core-set members are directly immersed in the vicissitudes of scientific practice, and it is their collective social and technical activity that results in the production of settlements on the facts of the matter under dispute. In most constructivist accounts of controversies, these settlements may then work their way into the wider scientific community and the public sphere, where they take on the status of truth as elements in the practice of routine science and common knowledge (Collins 1985; Latour 1987).

Understanding the social dynamics of core sets is important for making sense of how controversies end, but few constructivist studies of controversy have focused on the origins of controversies and the formation of core sets.[3] In many studies it is assumed that the most relevant actors in a controversy are either called upon by others or self-select by virtue of their interest or expertise with respect to the matter in question. It follows, however, that interests and expertise will vary depending on how the matter in question is perceived. Thus a specific actor's understanding of what it is that is being talked about (and who is already doing the talking) in part determines whether that actor may be drawn into a core set. At the broadest level this is an achingly simple point. If one happens to pick up an issue of *Science* and read an article, for example, entitled "Observation of Quantum Shock Waves Created with Ultra-Compressed Slow Light Pulses in a Bose-Einstein Condensate," it is unlikely that one will feel drawn into a discussion, let alone a scientific debate, if one is a lawyer or even a biologist. In this case the technical language of the article and the medium of its dissemination structure expectations for what the article is about. Even armed with basic skills for reading the article, one would still have to recognize what the problem was (if there was one) and then determine if one had any interest in pursuing the matter further.

Note that the interaction I am describing here is between a reader and a text (the article in *Science*). One may contact the authors directly or go to hear them speak at a colloquium, but unless one works in the same department, it is unlikely that the primary encounter will be with anything other than a text. In this sense, the text serves as a kind of invitation to further action that a reader may accept or decline depending on understanding and interest in what the text has to say. Normally, this invitation comes in the form of a peer-reviewed journal article or at least a technical communication of some kind, but with increasing frequency potential core-set members encounter controversies through more public and popular texts such as newspaper and television reports.

In the pages that follow, I will consider the effects of the first popular news media reports of cold fusion on the formation of the core set for the ensuing controversy. While there is no end of available methodologies for the study of scientific texts, my purposes are best suited by drawing on basic principles of the "socio-semiotics" of textual representation deployed by actor-network theorists in the sociology of scientific knowledge (Latour and Bastide 1986; Latour 1987; Law 1986). As John Law argues, words "operate to work upon the interests of the reader and force a 'route march' past a particular point that is held to be of key significance. . . . The scientific paper presents itself as an actor-world, an actor-world which operates upon the reader to translate him/her to a particular place in that world" (Law 1986, 68). Representation in this sense is treated as a form of strategic action meant to enroll the interests of audiences through a process of "translation," which configures the "identities of actors [and objects as actors], the possibility of interaction and the margins of maneuver" (Callon 1986, 203). The crucial idea, then, is not to focus on the desires and intentions of the authors of texts but rather on the relations between textual representations and readers as they jointly configure an understanding of what is going on. The point drawn from semiotics is that it is not the authors who are enrolling the interests of audiences (this would be the focus of rhetorical analysis) but the text, acting as a surrogate agent, that is doing the enrolling for the authors. As we shall see, this is important since understandings of a text may diverge wildly from authorial intention, and indeed authors' intentions may be exceedingly difficult if not impossible to identify, while the effects of texts on readers are not. With these basic principles in mind, it becomes possible to see how the early representations of cold fusion and their reception by diverse audiences provide a means of accounting for the formation of a core set of experts with specific kinds of expertise and expectations, whose social and material dynamics results in the disconfirmation of the phenomenon.

## *An Experiment That Changes Society*

"We are here today to consider the implications of a scientific experiment." With these words, Chase Peterson, the president of the University of Utah, began the press conference held at the Salt Lake City campus on the morning of March 23, 1989. The president was followed by the vice-president, who introduced two scientists to a room filled with university officials and members of the press. As noted in chapter 1, the two scientists were electrochemists: B. Stanley Pons was the chair of the chemistry department at the university, and Martin Fleischmann, Pons's one-time supervisor, was an emeritus professor visiting from the University of Southampton. Standing at the podium, Pons held up a small flask with some wires in it. "We have," he told the audience, "established a sustained nuclear fusion reaction by means which are considerably simpler than conventional techniques."[4]

The audience remained silent, as if not quite sure what had just been said. Pons continued by offering a brief synopsis of their discovery: "Deuterium which is a component of heavy water is driven into a metal rod exactly like the one I have in my hand here, to such an extent that fusion between these components—these deuterons in heavy water fuse to form a single new atom, and with this process there is a considerable release of energy and we've demonstrated that this can be sustained on its own. There is much more energy coming out than we're putting in."[5] Fleischmann went to the podium to add a few words, and then the reporters were invited to ask questions. In their answers, Pons and Fleischmann explained that they had observed a process akin to nuclear fusion but at room temperature, using an experimental procedure not much different from normal electrolysis.[6] The process produced neutrons, tritium, gamma radiation, and around four times as much energy (in the form of heat) as was added to keep the process going. They argued that more heat was being produced than could be explained in terms of a chemical reaction, and that the only other explanation would involve some nuclear process.

One reporter then asked what the social implications of their discovery might be. Pons opined that "it would be reasonable within a short number of years to build a fully operational device that could produce electric power." This seemed to catch the reporter's interest. Fleischmann's opinion was perhaps more cautious: "We don't know what the implications are, the subject has to be fully researched. . . . it is absolutely essential to establish the science base as widely as possible, as correctly as possible—to challenge our findings, to extend our findings. . . . but it does seem that there is here a possibility of realizing sustained fusion with a relatively inexpensive device." Then the vice-president of the university, James Brophy, stood up. Brophy made the important connection for the reporters in the room, providing them with the hook that would make this story newsworthy: "If indeed this scientific discovery proves to be as practical as it appears to be, not only does the world's population get a promise of virtually unlimited energy, it gets the elimination of acid rain, reduces the greenhouse effect and allows us to use fossil fuels in a way which is much more important than simply lighting a match to them." The next day, Fleischmann and Pons and the University of Utah made the front pages of newspapers around the world.

In principle, there was no reason why the press conference should not have been front-page news. The University of Utah was a reputable institution with one of the best chemistry departments in the country, and Fleischmann and Pons were acknowledged experts in their field. Further, as an exercise in marketing and public relations, the press conference would have to be seen as a resounding success. It established what a scientific paper could not, a connection or translation of a modest and incomplete set of electrolysis experiments to a future society in which energy would be cheap, clean, and safe. This translation was not lost on the media. By March 24, not only did reports from the press conference make the front pages of major newspapers around the world, but also the story became a lead item for television and radio news broadcasts. Fleischmann and Pons's announcement had clearly caught the interest and imagination of the popular media, and as a consequence their claims swiftly received an unprecedented amount of public attention.[7]

## *This Is Not Science*

Not all of the early attention given to the announcement of cold fusion was laudatory. The press conference prompted the immediate criticism if not anger of a number of scientists, which found expression in letters to journals and in editorials. Fleischmann and Pons's announcement had come as a surprise to the scientific community. Not only were most scientists completely unaware of the phenomenon Fleischmann and Pons were describing, but as was often pointed out in early news media interviews, there was no published article or even a less formal paper for scientists to refer to. By reporting their results in a press conference rather than through a peer-reviewed publication, Fleischmann and Pons had transgressed what many viewed as a norm of proper scientific conduct. In one angry article, for instance, Michael Heylin, the editor of *Chemical and Engineering News*, wrote that Fleischmann and Pons and the University of Utah had committed a "flagrant violation of the normal strict protocols for disseminating results of basic research" (Heylin 1989). In a manuscript written for a news magazine two days after the press conference, a prominent chemist at a Canadian university wrote, "The method chosen by Professors Pons and Fleischmann to disseminate their claims robs me and all other scientists of the opportunity to assess the details of these claims. . . . it is about time that errant members of the scientific community paid the price for teasing the public with news of dubious validity promising to better our lives profoundly."[8] In the view of this author, the information conveyed at the press conference and in the media was unscientific—a "tease" of "dubious validity." No real scientific judgment could be made on the basis of these reports.[9]

Yet announcing important scientific information in the media has become less unusual in recent years (Lewenstein 1992b), and as more research is conducted in private laboratories, the legalities of patent protection may override the desire for, or expectations of, formal and public peer review.[10] At one level, this seemed to be the case with Fleischmann and Pons. Fifty miles from Salt Lake City at Brigham Young University, another group of scientists led by Steven Jones was working on experiments that seemed to be similar to those of Fleischmann and Pons. Jones had been a reviewer for Pons and Fleischmann's Department of Energy grant proposal in 1988, and the two groups met to discuss what to do in early March 1989. Although the Utah and BYU teams reported very different results (Jones found neutrons but no excess heat), they had agreed to submit their work simultaneously to the journal *Nature* on March 24.

The exact details of what happened remain obscure.[11] In order to protect patent claims that the University of Utah had already filed in early March, university officials decided to hold the press conference as means of establishing priority for Fleischmann and Pons's work. Jones and his group viewed this as an abrogation of their agreement, and a fairly acrimonious priority dispute ensued. A press conference was held at Brigham Young University announcing Jones's results, and Jones began showing copies of his notebooks at scientific meetings, with details of his work dating back to 1986.

The press conference and representations of it in the media were important moments in the development of the controversy, but there are two different ways we can interpret that importance. For those who viewed the press conference as a transgression of norms of science, the question becomes what went wrong, and the door is opened for a sociological investigation of error. The argument here is that nothing scientific is happening, and what needs to be explained is the abnormal occurrence of the press conference and its representation in psychological or sociological, but not epistemic, terms. Yet whether it was perceived as a transgression or not, the press conference and its representations provided scientists and the public with a sense of what was going on in the laboratory at the University of Utah, and this opens the door to a true sociology of scientific knowledge and an investigation of the epistemic significance of the press conference.

Christopher Toumey (1996) argues that the representations in the media in the case of cold fusion can be "separated from the intellectual substance of science" and can take on a significance of their own as aspects of popular beliefs and ideologies. For Toumey, the case concerns the cultural symbolism of cold fusion as "the hope for a quick fix plus the simplicity of kitchen-table technology" (Toumey 1996, 121). In the end, an account of the actions and relations of scientists is not necessary, as they are seen to be distinct from the representations of those actions and relations in the media, which mix cultural symbols with scientific reality, creating a false image of what is going on.

A similar role for the media representation of cold fusion is suggested by Tom Gieryn (1992). Gieryn accounts for the life of cold fusion not through the interactions of scientists, who would have killed it off quickly had they worked alone, but through the production of a kind of mythology in the media. It is the "literary" replication, or the rendering of the experimental claim in narrative form, that promulgates and sustains a belief in cold fusion.[12] The force of this narrative is so powerful that it can sustain popular belief in cold fusion in spite of the presence of overwhelming scientific evidence against it. As Gieryn explains—and Toumey would no doubt concur—the narrative of cold fusion is about something more than what went on in a series of electrochemical experiments: "Out of the laboratory and into the world! Tell and retell a story about cold fusion. Give heart-warming history and a millennial future to associations forged among palladium, deuterium, neutrons, heat, calorimeters, Pons, Fleischmann, and the University of Utah. Write a biography that dresses up Pons and Fleischmann in something more fashionable, more heroic than white lab coats. Cast the moment of discovery as climax to an epic struggle, and replicate the story in newspapers and on television throughout the world" (Gieryn 1992, 223). Gieryn's analysis points to the ways cold fusion was configured in the media as a powerful society-transforming discovery, translating the interests of a vast array of actors (Latour 1983), allowing cold fusion to "live on relatively immune from disconfirming experiments" (Gieryn 1992, 222).

Gieryn is only partially correct about this; media representations did help facilitate the survival of cold fusion (as I will discuss below), but not by making it immune to disconfirming instances. The public narratives provided a context in which scientists attributed meaning to their experiments that could turn those experiments into disconfirming instances. The press conference and its representation in the media certainly involve something other than the making of a scientific argument, but what concerns us here is how these seemingly nonscientific "narrative" elements helped to shape and organize the practices of scientists and others who engaged directly in efforts to reproduce and validate the experimental claims. As I shall argue, contrary to Toumey, the rhetoric of the press conference and media imagery were not distinct from the intellectual substance of science but played an important role in configuring scientists' beliefs and practices with respect to what the phenomenon was and how it should be tested. For this reason, and in opposition to Gieryn, I will argue that the media shaping of cold fusion ultimately helped to kill it off, as well as enabling its continued survival.

## *Scientists' Response to the News*

Two things are worth noticing about scientists' initial responses to representations in the media. The first is that many certainly did question the credibility of information contained in reports in the popular media, but the second is that this concern did not prevent scientists from acting on these reports and rushing to their labs to try to reproduce the experimental effect that was being described. One electrochemist I interviewed, for instance, recalled his difficulty figuring out what was happening in the first weeks after the press conference: "All we had was a newspaper report. It wasn't enough to go on. We didn't even know what the electrolyte was. I knew that I could get in contact with Stan or Martin but it was really hard to do, everyone else was trying to do that and their phone lines were basically down or clogged all the time."[13] This kind of problem seemed endemic. Here, a physicist at Caltech expresses the same concern: "We certainly had tremendous difficulties trying to figure out what the people at the U of U had really done. Because everything we got had been filtered through the press."[14]

Clearly, scientists would have preferred a detailed technical report of some kind or a direct line of communication to Fleischmann and Pons, but the paucity of information did not keep them from working with what they could find. In an interview with Douglas Smith, Michael Sailor, a chemist who worked on cold fusion experiments at Caltech, recalls his experience of dealing with the media reports: "The *Financial Times* told of this new boundless source of nuclear-fusion energy obtainable at room temperature. Bob Finn, in our Public Relations office, got me a copy of the AP press release based on the *Times* article almost immediately. It had no details, so we tuned into every newscast we could all day, trying to decipher what Pons and Fleischmann were really claiming to have done. And we speculated on how it might work. . . . We immediately realized that we could probably do a better job on the neutrons than almost anybody else in the world."[15] From Sailor's account we can begin to see how the apparent lack of detail in the media was supplemented by the expertise of scientists who were able to speculate and make decisions about what might be happening. Sailor was not alone. Within hours of seeing the news broadcasts or reading the newspaper articles about the press conference, hundreds of scientists around the world dropped what they were doing and headed to their labs to see if they could reproduce the effects that had been reported.[16]

While it makes sense that scientists and nonscientists would interpret media representations of cold fusion differently, comments like those above from scientists indicate that such representations were hardly ignored or dismissed, albeit considered less than fully credible. Even scientists who were initially skeptical of Fleischmann and Pons's claims based their skepticism on the information they had gleaned from newspapers. In the unpublished manuscript criticizing the press conference quoted above, the Canadian chemist also made the following argument based on information that was conveyed in the media: "The trouble is there are enormous energy barriers to surmount before atoms can be fused. . . . For fusion to occur the two deuterium atoms must be brought together ten thousand times more closely than they normally like to be. There are simply no nooks or corners in palladium that are both so tiny and so strong to act as the die in which helium could be forged. The claims of Professors Pons and Fleischmann are therefore almost certainly wrong."[17] This criticism of cold fusion became common enough as the controversy progressed, but here the author treats the newspaper reports he has read as being reliable enough to make a judgment about the validity of Fleischmann and Pons's claims, even though he has not yet seen their technical paper. Scientists were not simply responding to cultural symbolism and mythological narrative; they were engaging with media representations as legitimate and enabling scientific accounts. This was in spite of the moral outrage shown by those scientists and commentators who perceived the press conference announcement as a breach of proper scientific conduct.

The comments of the scientists noted above show that while formal peer-reviewed publication serves as an ideal of proper scientific practice, it is not necessary to that practice. Indeed, given the reward system in science, a high degree of credibility in cases of novel experimental claims may not be desirable. Collins (1975) has argued, for instance, that in cases of anomalous experimental claims or extraordinary science, few scientists are actually initially interested in replicating experiments precisely, since the potential rewards for such work (in terms of reputation, awards, or money) would not be worth the effort. What scientists may seek to do in such cases is to build an apparatus of their own design in an effort to produce the original results, but also to "go one better" by adding to the knowledge base.[18] In this regard, the greater the credibility of an experimental report, the less scope there might be for scientists to vary their experiments and interpretations of it. The uncertainty associated with media reports arguably leaves scientists with more room to maneuver, and this enables them to chart their own, possibly better, course toward producing the phenomenon. Thus, one engineer I spoke with comments on his memory of reading about cold fusion for the first time and explains: "So what I thought was that it seemed that in their system there is something that was not completely clear so let's have a look and see what it might be. I had no idea of a theory or what it could be but as I thought it might be interesting to generate heat I thought it was interesting to go and see experimentally what happened. . . . I had the idea . . . to replace the water used, the heavy water used by Fleischmann and Pons, by gaseous hydrogen."[19] The lack of technical clarity in vague newspaper reports facilitated this scientist's development of an alternative method for producing the effect. In this case the reader is confronted with an incomplete text. The lack of detail combined with the mythological character of the cold fusion claim becomes a kind of puzzle that invites certain readers to participate and find a solution. But what kinds of readers are invited by these texts? And what sorts of solutions do they propose?

Lewenstein (1992a, 1995) is one science-studies scholar who has already pointed out the ways in which scientists drew upon news media reports in formulating opinions about the credibility of the cold fusion claims as well as how to test them. He has also shown how some reporters became intimately involved in scientific decision making by acting as conduits of information between different scientists doing cold fusion work. Lewenstein (1995) argues that media involvement in aspects of scientific practice served to confuse or delay the formation of a scientific consensus because of the multiplication and complexity of the number of "channels of communication" involved.

In Lewenstein's account, media reportage mixed and conflicted with other kinds of representations, making any collective resolution of the status of cold fusion difficult and problematic. Once the news media began to lose interest, however, the communication of information stabilized, and so did knowledge about the phenomenon. Like Gieryn, Lewenstein suggests that the confusion generated by the media helped keep cold fusion alive by keeping the controversy open. My own interpretation differs in that cold fusion as an experimental object, and the core set of researchers that engaged in practices aimed at confirming its existence, were in part configured, and not confused, by multiple representations in the media. As the creators of the most accessible and widely disseminated kinds of representations, the role of the media in facilitating certain kinds of experiments and interpretations that required certain kinds of experts (i.e., core-set members) rather than others needs to be accounted for.

## A Public Scientific Event

Communication in science is normally a fairly slow, highly mediated process. Most scientific information is never circulated beyond a small, very specialized community of experts. Because of this, scientists are usually aware of only a tiny fraction of what other scientists are doing, and this is especially the case if they work in different fields. In the rare instances that information does move into the public sphere, it is thought to do so incrementally as communication proceeds from individual experts through conferences and journals to scientific news media, and then to the popular news media.[20] Under these conditions, most specialists in a scientific field know about the latest information before members of the lay public. This was not the case with the announcement of cold fusion. Fleischmann and Pons, in consultation with university officials and lawyers, decided to hold a public press conference, preempting the publication of their observations in the *Journal of Electroanalytical Chemistry* by about two weeks. In addition, the two scientists did not discuss their work with many colleagues, so that when experts in the fields of electrochemistry and nuclear physics read about cold fusion in the newspaper on March 24, it was, for most, the first time they had heard about the work.

Thus the pattern of communication was the exact reverse of what many expect as the norm. For the first week after the press conference, information was available to scientists only in newspapers and television and radio broadcasts. Excerpts from the press conferences (the University of Utah event on March 23 and the Brigham Young University event on March 24) were played and replayed on the evening news. On March 26, Stanley Pons was interviewed on the *Wall Street Journal Report*, advising that scientists shouldn't try to repeat their experiments until a published paper was available, but that did not stop the many researchers who went to their labs the moment after they had heard the news on March 23 (Bishop 1989b). Then, on March 28, Martin Fleischmann gave a more technical presentation on their experiments at the Harwell Nuclear Research Laboratory in England, and notes from that event began to circulate on the Internet, where talk of cold fusion had already become the central topic of science-oriented USENET newsgroups and electronic mailing lists. Interestingly, while the occasional message like ones reporting on the Harwell talk would appear on the Internet, the most prominent function of electronic lists in the early days of the controversy was in fact the circulation and distribution of newspaper reports from around the world.

By the end of March, however, information from newspaper media was beginning to mix with other more traditionally credible sources. For a variety of reasons, few core-set researchers seemed to make much use of electronically distributed messages other than private e-mail they might receive from friends, but during the first week of April a draft preprint copy of a paper authored by Fleischmann and Pons was making the rounds of major research laboratories via fax, and copies of this unauthorized paper were treated like gold. By the time major science news circulars like the journals *Nature* and *Science* and magazines like *Chemical and Engineering News* published more in-depth stories attempting to provide broader background, cold fusion already had a significant worldwide public presence. These stories were followed by the formal publication of Fleischmann and Pons's preliminary report (a final version of the faxed preprint) in the *Journal of Electroanalytical Chemistry* on April 10, and an article by Jones's research group in *Nature* on April 27.

While the Utah announcement and the rush to validate the cold fusion claims were, in their own right, a tremendous scientific event, it is important to point out that their presence in the general media made that event public on a massive scale. Here, I do not want to suggest only that the cold fusion controversy had a large public audience in the way, for instance, that the public constituted an audience for the NASA space program in the late 1960s.[21] Rather, the cold fusion controversy should be viewed as a public scientific event in which a public composed of more than just a small group of experts becomes integral to the practice of science. A number of basic observations about the role of the media in this case will help to make this distinction clear.

First, we must be attentive to the scale on which communication about cold fusion took place. Prior to March 23, only a small handful of scientists, administrators, and bureaucrats knew of Pons and Fleischmann's work; after March 23, that number increased by thousands, if not tens of thousands. While most readers were nonscientists with little more than a passing interest in the story, the scientific audience generated by the news media attention far exceeded any that might be available for a report published in even the most popular scientific journals.

Second, not only were reports of the existence of cold fusion accessible to thousands of people; they were accessible to all of those people within a very short period of time and, crucially, at around the same time (within a week). This is a marked contrast with the rate at which information is normally disseminated via scientific journals or even through more informal networks of communication. In these cases, the rates of dissemination are much slower, and the information being passed along is increasingly mediated at each stage (as an article passes through an editor, peer review, and revisions, or as scientists talk with one another over coffee at a conference). In the case of cold fusion, scientists all over the world became aware of the phenomenon, subject to about the same degree of mediation, at the same time.

Third, the scope of even the scientific audience for media representations of cold fusion was vast, and as a consequence so was the scope for participation in efforts to reproduce the phenomenon. This is a crucial point, as the conditions for the composition of the core set are radically altered. Journal articles have small scientific audiences not only because of their limited distribution, but also because of the language in which they are published and the use of specialized terminology. The limited audience reduces the pool of actors from which a core set will be constituted. Worldwide media coverage meant not only that scientists in many non-Western and non-English-speaking countries knew about cold fusion, but also researchers from a wide variety of different fields. This larger and more varied scientific audience becomes the actual pool from which core-set participants in the controversy may emerge.

Fourth, the media reportage on cold fusion was also accessible to a significant population of other kinds of actors who complete the complex dynamic of science worlds.[22] These include nonprofessional scientists or unaffiliated researchers like graduate and undergraduate students, retired and semiretired scientists, managers and administrators, amateurs, technicians, and so forth. There are also groups of scientific support personnel, actors who do not themselves engage in research but help facilitate it: politicians, corporate executives and university administrators, investors and entrepreneurs, equipment manufacturers and suppliers, as well as science writers, critics, and some sociologists of science.

From these observations, one can begin to get a sense of the role of the media and their representations in shaping the epistemological features of the controversy in terms of the potential audience of actors that could be drawn into the dispute. Starting from such a large pool of potential participants, it remains to be seen how media representations drew in or invited some notions of expertise (and some experts) rather than others. For this we must look to how the phenomenon of cold fusion was configured in the first media reports.

## *Configuring Cold Fusion*

The following is the headline from the first page of the *Wall Street Journal* for March 24, 1989:

Taming H-Bombs?

Utah Scientists Claim Breakthrough in Quest for Fusion Energy

---

If Verified, Their Experiment Promises to
Point the Way to a Vast Source of Power

---

Batteries and Palladium Wire

Headlines like this not only provide the first representations, and therefore the first glimpse, of the new phenomenon, but they also simultaneously establish a framework for both lay and scientific understandings of it. Here, the same representation seems to configure the phenomenon in question (as something related to "Batteries and Palladium Wire") as well as articulate its broader cultural significance (as in "Taming H-Bombs" for use as a "Vast Source of Power"). As Toumey and Gieryn have argued, this kind of representation generates mythic narratives, but it also offers statements about an experimental reality that could be practically engaged.

In the article published under the headline above, Jerry Bishop and Ken Wells provide this account of what was said at the press conference: "The scientists said that with no more equipment than might be used in a freshman chemistry class, they had triggered a fusion reaction in a test tube that continued for more than 100 hours. The reaction, they asserted, passed the so-called break-even point, producing more energy than was needed to trigger it. An experiment now in operation is producing four watts of power for each watt of input" (Bishop and Wells 1989). The key feature of this account and indeed all the news reports of the Utah experiments was that what was going on was some kind of fusion reaction that produced more energy than it consumed. The scientists' confidence in fusion as an explanation for what was going on is presented as a robust fact supported by the seeming simplicity of the experiment ("freshman chemistry class"), the duration of the effect ("more than 100 hours") and its magnitude ("four watts of power for each watt of input").

Both the account of the press conference and the image of the experimental object generated within this text act to configure the phenomenon in question as a particular kind of object with a recognizable cultural and technical history: the history of nuclear fusion. This is not inherent in the object itself, since no one is absolutely sure what it is yet, not even Fleischmann and Pons. Nor does this media story simply reproduce Fleischmann and Pons's account of the phenomenon. Rather, the object and its history are products of their representation. The moment it was reported in the media was the moment the entity under investigation became a kind of nuclear fusion (soon to be designated by the term "cold fusion") for both the lay public and the majority of scientists who would become involved in the controversy.

Did Bishop misrepresent Fleischmann and Pons? I want to argue that this is not the issue, although numerous scientists on both sides of the controversy have lamented the use of the term "fusion" in describing the Utah experiments. In a BBC interview in 1997, Fleischmann claimed that in 1989 "we did not actually say we had achieved fusion. We said we had created large amounts of energy which could not be explained by chemistry" (Tinsley 1997). But Bishop and other reporters did not arbitrarily assign the term. The phenomenon was referred to as a form of nuclear fusion in the University of Utah press release and at the press conference, particularly in the comments made by James Brophy, the vice-president of the university.[23] Fleischmann and Pons also did not deny (or did not successfully deny) that fusion was an apt explanation for their experimental results. Indeed, at an American Chemistry Society meeting on April 12, 1989, Pons referred to the electrolytic cell as the U–1 Tokamak, a tongue-in-cheek reference to the American nuclear fusion research program. For better or for worse, then, the media representation of cold fusion as "fusion" was a collective construction.

This collective construction occurred in at least four different representational contexts that appear in various newspaper and media reports: a cultural context, a pedagogical context, a disciplinary context, and an evidential context. These contexts act like signposts; they signal the presence of a phenomenon that both draws the attention of experts in nuclear physics (instead of experts in biology, for instance) and directs their line of interpretation in terms of what the phenomenon might look like, how it might behave, and so on.

## THE CULTURAL CONTEXT OF FUSION

The term "fusion" has a particular technical meaning as well as a broader cultural signification. Technically, fusion refers to the physical reaction that occurs when the nuclei of two or more atoms combine. Culturally, fusion represents the apotheosis of energy research in the twentieth century. It is this cultural significance that generates the media-worthiness of the cold fusion story. On the evening of March 23, the story of the press conference was featured on the television network news, accompanied by poignant visuals of Fleischmann and Pons and their experiments. On one broadcast, the image of an electrochemical cell was juxtaposed with an image of the sun in an exchange of visual signification, along with the following narration: "It's what keeps the sun burning—atoms fusing together. It's what powers the hydrogen bomb. If you could harness it you could have a virtually limitless supply of energy. Today scientists Martin Fleischmann and Stan Pons told a news conference at the University of Utah that they had done just that in a test tube."[24] The *Wall Street Journal* article quoted above exclaims, "If verified, their experiment promises to point the way to a vast source of power." This statement is followed by already recognizable hyperbole about nuclear fusion: "One gram of deuterium-tritium fuel, about three hundredths of an ounce, could produce as much energy as 2,400 gallons of petroleum"; a similar statement appeared in the *Financial Times*: "If their discovery is confirmed, they will have gone a long way towards taming the forces powering the sun and the hydrogen bomb. These could provide virtually unlimited, clean and inexpensive energy" (Cookson 1989).

Statements like these form the basis of Toumey's argument that cold fusion is "conjured science," but scientists doing experiments were audiences for them. The technical and cultural meanings of fusion combine in the media to make cold fusion important and interesting to scientists and nonscientists alike. As one scientist commented, "This is potentially the most important experiment of the twentieth century. And I've been an electrochemist for a decade, and all of a sudden electrochemistry is the most important issue in science. And I can make a contribution. My god, it has fallen right into our laps."[25]

This rhetoric is not new. It has been a steady feature of reporting on, as well as lobbying for, nuclear energy in North America and Europe since the 1970s. Thus the conceptual framework for presenting and understanding cold fusion in terms of a world-changing energy revolution already existed. Lievrouw (1990, 6) refers to this as a process of "anchoring" the concept of cold fusion in "the existing public perception of fusion energy." In this process, Fleischmann and Pons's phenomenon became attached to public discourses of nuclear fusion, and audiences were invited to view cold fusion in terms of this discourse.

### THE PEDAGOGICAL CONTEXT OF FUSION

Almost all the early media reporting on cold fusion featured a one-or-two-paragraph lesson in the physics of nuclear fusion and fission (usually accompanied by a diagram). In the context of the overall articles, these physics lessons situate cold fusion as part of a relevant scientific domain. The reader is told that fusion, as opposed to fission, involves the combining or "fusing" of hydrogen nuclei but that this is difficult because hydrogen nuclei have a positive charge and ordinarily repel one another. To overcome this repulsion and generate fusion, large amounts of energy are normally required.

For lay readers, this information situates Fleischmann and Pons's announcement as a "discovery" and gives them a way of making sense of both the implied technical achievement and its unconventional character. Fleischmann and Pons are presented as having seemingly gone beyond textbook physics, opening up a new set of scientific problems. What these physics lessons do for expert audiences, however, is to reify what is unconventional about cold fusion. Cold fusion comes to be defined in terms of a break with the standard physics represented by the lessons on fusion. There are two important consequences of this: the first is that in defining cold fusion this way, the newspaper text identifies nuclear physicists as the most relevant experts in the case, and the second is that it simultaneously precludes alternative readings of cold fusion consistent with standard physics.

### THE DISCIPLINARY CONTEXT OF FUSION

"Cold fusion" was not the term initially used by Fleischmann and Pons to describe their research. Rather the term is a product of a conflation in the media between Fleischmann and Pons's experiments and the simultaneous work of Steven Jones. Indeed it is appropriated from the work of Jones. Aside from the coincidence of the announcements linked to the priority dispute I described earlier, Jones's public announcement helped make cold fusion into a pressing scientific problem for physicists.

Jerry Bishop's initial article in the *Wall Street Journal* was the first context in which "cold fusion" was used to describe the work of Jones and Fleischmann and Pons: "The Utah research, which until now has been carried out quietly without any publicity, is believed to follow an approach considered a dark horse in the race to hydrogen fusion. Known as catalyzed or 'cold' fusion, it's an approach being attempted not only at the University of Utah in Salt Lake City but also at Brigham Young University in Provo, Utah" (Bishop 1989a). The mixing of the two sets of experiments in this public context produced the hybrid "cold fusion," which has since become the popular term for the controversial phenomenon.[26] The kind of fusion Fleischmann and Pons were postulating may have seemed unlikely as far as most physicists were concerned, but Jones's experiments provided physicists with an "in," a way to see the Utah results in terms of an established tradition of research in conventional nuclear physics. Unlike Fleischmann and Pons, whose expertise in nuclear physics could be questioned, Jones was considered an expert in a relatively obscure but acknowledged subfield of high-energy physics conventionally understood as muon-catalyzed fusion, or cold nuclear fusion.[27] Jones's expertise in the field of muon-catalyzed fusion was undisputed. His work at BYU, which led up to the March 24 press conference (and a subsequent paper published in *Nature*), was part of an investigation he and his colleagues had first articulated in 1978.[28]

In collaboration with a number of fellow scientists and graduate students, Jones designed a set of electrolysis experiments using palladium and titanium as cathodes and an electrolyte composed of heavy water and a complex mixture of metal salts that came to be known as Jones's "mother earth soup." Although the BYU researchers did not initially attempt to measure excess heat, they had a highly sensitive neutron spectrometer designed to detect very low levels of neutrons. What they reported was the detection of neutron fluxes only slightly above background, many orders of magnitude less than what was reported by Fleischmann and Pons. The priority dispute that evolved between Jones and Fleischmann and Pons centered on who had performed electrolysis work with palladium and deuterium first, even though the groups did not claim to be observing the same kind of effects. While the dispute itself eventually became a mute issue as the controversy began to turn against cold fusion altogether (both the University of Utah and the BYU version), the fact that the dispute occurred at all suggested that Jones on one side and Pons and Fleischmann on the other had run across similar or related phenomena.

The way in which the Utah and BYU claims were conflated can be seen in the comments of one physicist, who had this memory of learning about cold fusion:

> I was lying in bed on Good Friday and I went out and bought the *Globe and Mail* and I came back, and I was lying in bed next to my

> wife and on the front page of the *Globe and Mail*, on the bottom half, there's this picture of a electrolysis cell, which catches my eye, and I started reading this article and I hear about Pons and Fleischmann's press report. So I'm just sort of lying there and I start thinking about it, because I'm a high-energy physicist. The nuclear physics, I would presume to be an expert on, but I said this is so bizarre that if it's true there is a good chance that it's got something to do with what I'd call fundamental physics. So I was lying there going though my head thinking what could it be, cause I know about muon-catalyzed fusion and things like that so I was thinking about this—what kind of particle could it be, what kind of binding it has, what kind of couplings it has.[29]

Here, the newspaper report generates a conception of cold fusion as being similar to muon-catalyzed fusion, as well as drawing in the physicist as someone with expertise appropriate to understanding the phenomenon. In this case, the physicist went on to perform a series of cold fusion experiments and later joined other physicists in arguing against the reality of cold fusion.

A number of commentators have pointed to the discrepancy between the results reported by Jones and Fleischmann and Pons as a reason to be skeptical of cold fusion, but the opposite was the case.[30] The publicity surrounding the priority dispute suggested that the BYU experiments constituted some kind of independent replication of the phenomenon that Fleischmann and Pons had reported (or vice versa). Initially at least, Jones's results added to the credibility of Fleischmann and Pons's claims, not only in the form of an independent replication, but also in the form of a respectable research pedigree in nuclear physics. In this way, public representations invited not only the interest of nuclear physicists, but also provided them with a way to investigate the anomaly, as a form of conventional fusion.

### THE EVIDENTIAL CONTEXT OF FUSION

Finally, for scientific audiences and experts in nuclear physics, simply calling the object fusion might not be enough to draw their interest and participation, especially if a reporter, a university official, or even a chemist is making the claim. But there is more going on in the text than simply naming the phenomenon "cold fusion." There are also observational statements that collectively make up what Pinch has called the "evidential context" of the observation report (Pinch 1985, 10–11). It is the evidential context of media reporting on cold fusion that makes the object significant as a form of nuclear fusion.

According to the *Wall Street Journal* article of March 24, the cold fusion reaction "continued for more than 100 hours" and "passed the so-called break-even point, producing more energy than was needed to trigger it," and later the reader finds that the experiment involves "a test tube of heavy water" and that "deuterium nuclei are brought close enough together to overcome their mutual repulsion and then fuse." Further, the article cites the detection of "neutrons, tritium and helium—the expected by-products of fusion reactions." All of these elements are immediately identifiable by scientists (even those who are not nuclear physicists) as being evidence that the object in question is a kind of nuclear fusion.

The newspaper reports thus up the evidential ante by guiding an interpretation of the Utah experiments in terms of conventional fusion. Again, one can certainly look at this process as a second-order effect of the rhetorical construction of cold fusion in the press conference. In this sense we could view Fleischmann and Pons (and University of Utah officials) as taking an unjustified risk by rhetorically increasing the evidential significance of the data in the hope of garnering support and resources for their work.[31] But the crucial difference between the press conference and media reports is that Fleischmann and Pons are not there to mediate the flow of information. They are not able to expand on a point, deny what is being said, or even obfuscate—this task is left to the reader. If readers are going to respond at all to the newspaper report, then they are being called on by the text to interpret "cold fusion" in terms of nuclear fusion. In the hands of experts wishing to do experiments, this interpretation leads to the design of replication attempts that can then be evaluated on the basis of the presence or absence of recognizable signs of a fusion reaction.

## *Cold Fusion in the Scientific Literature*

I have thus far considered the ways in which public representations in the media in the first week after the press conference configured cold fusion as a phenomenon understandable and therefore criticizable in terms of fusion physics. Amongst other things, the nomenclature used to describe the Utah claims points to the role of newspaper reports in inviting potential experts to view the situation as a nuclear physics kind of problem. I want to extend this analysis by looking more closely now at the scientific literature, specifically the first paper authored by Fleischmann and Pons (along with Pons's graduate student Marvin Hawkins) (Fleischmann, Pons, and Hawkins 1989a). Media and public representations configured cold fusion in a way that drew in large and diverse audiences, who then became readers of the Fleischmann, Pons, and Hawkins paper and active core-set members. For a month or more the newspaper reports and this article were the only widely available textual accounts available to researchers.

My suggestion here is that the argument and structure of this scientific text, combined with the media, configured expectations of readers to help produce cold fusion as an anomaly in nuclear physics, and that this understanding directed subsequent interpretations of the various attempts at replication. Again I am less concerned with understanding the motivations of the authors (in this case Fleischmann and Pons) than I am with making sense of how core-set members direct their activities based on interpretations of the texts with which they are presented. With this in mind, it is interesting to note that the Fleischmann and Pons paper, like their press conference announcement, did not appear in the manner of most scientific publications.

As mentioned above, within a week of the initial press conference and nearly two weeks before formal publication, preprint copies of the article became widely available by fax. Prior to the press conference, Fleischmann and Pons had submitted this article to the *Journal of Electroanalytical Chemistry* (*JEAC*), a major journal for electrochemistry. The journal's American editor, Ron Fawcett, agreed to publish the article as a preliminary note in order to help establish priority for the work.[32] The paper appeared in the April 10, 1989, issue of *JEAC*, and was distributed at the annual American Chemistry Society (ACS) meeting during a special session on cold fusion. In the following issue of the journal, Fleischmann and Pons published errata for their paper (Fleischmann, Pons, and Hawkins 1989b). This made matters somewhat complex; a researcher might have a copy of the article but not the errata, or just the preprint and not the final article. While the differences in the various versions were slight (some changes in wording and in a couple of the calculations), the different circulating versions reinforced the sense of unprofessionalism around the work that had already been articulated with respect to the press conference.

The reception accorded by scientists to the *JEAC* paper was similar to the reception of the initial media reports. Electrochemists and physicists alike complained that the paper was "sloppy" and "uninformative." "There were no details on analysis or controls."[33] One comment written by a scientist in the margins of a preprint I obtained read, "Awfully incomplete after 3-1/2 years of work." Another physicist told me, "If people had the paper in their hands right from day one, I don't think it would have gone so far so fast, because that paper from my perspective is really bad."[34]

Of course not all scientists were this critical; like the newspaper reports, the paucity of information in the *JEAC* paper facilitated the possibility of innovation. Within weeks of the publication of the paper, scientists at MIT and elsewhere were already filing patent applications based on alternative approaches. One Japanese electrochemist, Tadahiko Mizuno, wrote a book about his experience with cold fusion, including the following account of his first reactions to the story in a newspaper on March 24:

> Because of my extensive experience with electrolytic systems, the news of cold fusion made a big impact on me. March 25, the day after I heard the news, was a Saturday. I thought I would start a replication experiment that very day. . . . The palladium cathode was the problem. I had obtained a fax copy of the Fleischmann-Pons paper but it did not describe any details about the materials or the electrolyte. I had no idea what the proper configuration should be, or the purity of the materials, or what material treatments to perform. Based on my experience, I concluded the main goal should be to load as much deuterium into the metal as possible. (Mizuno 1998, 36)

Again, it appears that researchers who began cold fusion experiments found little more information in the scientific paper than what was already present in media reports, but the lack of detail did not stop them.

The appearance of the *JEAC* paper, no matter how uninformative, did have an important impact on the trajectory of the controversy. Unlike media reports, Fleischmann and Pons could be held accountable for the statements in the paper. Where critics of cold fusion may have excused media representations as distortions or bias, statements made in the *JEAC* paper could more readily be taken as credible statements that one could agree or disagree with. For this reason the technical issues of the controversy tended to turn around statements made in the paper. Indeed, that the main issues of the controversy should turn around statements in the *JEAC* paper was itself a matter of dispute, as we shall see in the next chapter. In what follows I will briefly outline the main arguments of the paper and consider the ways in which certain statements helped direct the trajectory of the controversy.

## *The Elements of Controversy*

As they tell the tale in the *JEAC* paper, Fleischmann and Pons took their cue from earlier work in the physical properties of metal hydrides. They designed their experiments to test the hypothesis that the fusion of deuterium dissolved in palladium might be possible. Their argument rests on the assumption that deuterons inside the metal are in a highly compressed and mobile state, thus significantly increasing the likelihood for fusion reactions to occur. To test the hypothesis, Fleischmann, Pons, and Hawkins constructed a kind of electrolysis experiment.

Their apparatus consisted of a Dewar flask suspended in a water bath fitted with the standard components for conducting electrolysis experiments. In this case, however, palladium was used for a cathode (in the form of a thin rod anywhere from 1.5 cm to 10 cm in length) and platinum wire for the anode. The cathode and anode were set inside the Dewar filled with a solution of heavy water ($D_2O$) and lithium deuteroxide (LiOD). When a controlled current was applied to the cell, the heavy water electrolyzed, reducing the molecules to ionic deuterium and oxygen. The oxygen ions migrated toward the positively charged anode and bubbled out of solution as oxygen gas, and the deuterium ions migrated toward the palladium cathode and were absorbed into the metal. Under constant electrostatic pressure, the deuterons might remain dissolved in the metal. If the current were shut off or lowered past a certain point, the deuterium would begin to bleed or bubble out. The main purpose of the apparatus, then, is to generate the conditions under which high concentrations of deuterium can be maintained inside the cathode. The variable elements of the experiment include the size and shape of the cathodes, the current density, and the charging time (or duration that the cathode is subject to electrostatic pressure).

The reaction that takes place inside the palladium cathode is meant to be the phenomenon under investigation. In their paper, Fleischmann, Pons, and Hawkins's indirect evidence for nuclear fusion is obtained by measuring what are conventionally thought to be the products of fusion: energy in the form of heat (or enthalpy) and charged particles. The energy production is measured using "static" calorimetric techniques to determine the relative degree of heating of the electrolyte and the water bath in which the electrolytic cell is immersed (using thermistors placed inside the flask and in the water bath).[35] Fleischmann, Pons, and Hawkins also measured three kinds of charged particles. The emission of gamma rays was measured by a scintillation detector, the emission of neutrons was measured by a neutron detector, and the rate of tritium production was measured by analyzing samples of the electrolytic solution taken at various intervals.

The presentation of experimental data was divided into two sections: calorimetric measurements and measurements of charged particles. The two sets of data together form the basis of the claim for fusion, although greater emphasis is placed on the heat measurements. While a range of excess heat data is provided (expressed as a percentage of break-even),[36] it is on the basis of the data from only a few cells that the following claim is made: "enthalpy generation can exceed 10 W cm–3 of the palladium electrode; this is maintained for experiment times in excess of 120h, during which typically heat in excess of 4 MJ cm–3 of electrode volume was liberated. It is inconceivable that this could be due to anything but nuclear processes" (Fleischmann, Pons, and Hawkins 1989a, 307).

The force of this claim stems from the fact that chemical reactions, which are reactions at the molecular level, involve energies lower than those of nuclear reactions (reactions at the atomic and subatomic level). In this case, the implication is that if one adds up all the chemical reactions that would be taking place inside the electrolytic cell, the energies released still could not account for the measured excess heat.[37] By process of elimination, some nuclear process must be involved. In the *JEAC* paper readers were asked to accept this conclusion on trust since no accounting of the sources of heat is provided. This "energy account" later became a major source of controversy.

To help support the claim of anomalous heat, Fleischmann, Pons, and Hawkins provided data based on several calculated projections using larger cathodes and higher current densities. These data were often referred to in the media, but they are projections and not the results of actual experiments; this generated some serious confusion between actual and anticipated effects. One table in the paper, for instance, lists projections of excess energy of over 1,000% of break-even. There is no doubt that such projections can be read as either exciting or preposterous (depending on one's point of view), but as numerous readers pointed out, they are hypothetical, and it is a matter of some dispute as to whether the effects of the actual experiments could be scaled as Fleischmann and Pons report.

Lastly, Fleischmann, Pons, and Hawkins reported one event in which part of a cathode was vaporized and the cell was destroyed, presumably due to the generation of large amounts of energy. This event became an important piece of anecdotal evidence for the reality of cold fusion since, paraphrasing Hacking (1983), if the phenomenon can melt palladium then it must be real. Fleischmann and Pons also made use of this piece of evidence to suggest that the cold fusion experiments had the potential to be extremely dangerous, producing energy in an unpredictable and possibly explosive manner. Fleischmann later intimated that this observation was linked to his feeling that cold fusion might have a military application and that he was keenly worried about this prospect.

Much less attention was paid to the presentation of charged-particle data (neutrons, gamma rays, and tritium), but for nuclear physicists this constituted the most important evidence for the claimed effect. Fleischmann and Pons, like Steven Jones, reported that significant radiation was being produced in their cells but that it was far less than would be expected given the measurements of excess heat in terms of conventional fusion processes.

With respect to this there are three important claims. First, the experiment produced gamma rays at an energy level that confirms the production of 2.45 MeV neutrons in the cell. This is crucial evidence since 2.45 MeV neutrons are caused by a known fusion reaction and cannot be attributed to background radiation. For many readers this became the most convincing evidence that some kind of nuclear reaction was taking place in Fleischmann and Pons's cells. Second, there was a flux of around 4,000 neutrons per second in one experiment, a rate three times higher than the normal background neutron count. But while this count appeared to be higher than background, it was also much lower than would be expected if fusion were occurring by the same reaction that produced the gamma rays or excess heat. At the same time, however, the presence of above-background levels of neutrons is a potential indicator of a fusion reaction. Taken with the other data (including the independent neutron measurements of Steven Jones), the neutron results seemed to support an anomalous-fusion hypothesis. Third, Fleischmann, Pons, and Hawkins detected small concentrations of tritium in one cell that had produced excess heat. The concentration reported was extremely small, but blank experiments run with platinum electrodes rather than palladium produced almost no tritium, and this suggested that the tritium was more likely due to a nuclear reaction than to contamination or impurities in the electrolyte.

Drawing all these results together, the *JEAC* paper argued that, given the observed anomaly of large amounts of excess energy coupled with the emission of small amounts of neutrons and other charged particles, the conventional models for describing fusion reactions could account for only "a small part of the overall reaction scheme and that the bulk of the energy release is due to a hitherto unknown nuclear process or processes" (Fleischmann, Pons, and Hawkins 1989a, 308). This claim, lacking any conditionality (Latour 1987, 44), became the focal point for the developing controversy. It extends beyond the observation of mere anomaly to suggest that conventional theories of fusion, and by implication nuclear physics and quantum mechanics, may be incomplete or incorrect. To explain why this might be, I will need to examine conventional theory in more detail.

## Cold Fusion as a Nuclear Physics Problem

Loosely speaking, the study of nuclear fusion and also fission arose from the theoretical and practical exploitation of the relation between matter and energy expressed in Einstein's famous equation $E=mc^2$. In the case of nuclear fusion, the nuclei of light atoms collide, forming slightly heavier nuclei.[38] The new nucleus has a mass that is less than the combined masses of the fusing nuclei, and the difference in mass is released as energy. The exemplary case for understanding fusion processes is the Sun, which heats Earth by converting hydrogen to helium, producing energy in the process through a complex chain of nuclear reactions. Fusion can occur only if nuclei have enough kinetic energy to overcome the repulsion between them, and this is why extremely high temperatures and pressures are required to initiate any kind of fusion reaction at all. Such conditions are usually found only in the state of matter known as a plasma. The probability of fusion taking place between any given nuclei is known as the fusion cross section for that reaction, and for the most part the probabilities are extremely low to nil on Earth (which, of course, is a good thing for us; otherwise the Earth might burn up like the Sun).

The fusion process is complex, with many intervening steps (collectively called a reaction chain) that vary depending on what kinds of particles are fusing. Reactions between nuclei are conventionally described in terms of their "branching ratios," or the relative probabilities that certain fusions will produce certain products. These branching ratios can be theoretically calculated and experimentally verified, as they have been for the most common fusion reactions. The branching ratio for the reaction between two deuterons, or D-D fusion, is the major point of contention in the controversy over cold fusion and is expressed as follows:

(1) D+D → triton (1.01 MeV) + proton (3.02 MeV) + energy;
(2) D+D → helium–3 (0.82 MeV) + neutron (2.45 MeV) + energy;
(3) D+D → helium–4 + gamma rays (23.77 MeV) + energy.

Reactions (1) and (2) occur about 50% of the time, while reaction (3) is extremely rare, happening only about 0.00001% of the time. In reaction (1) deuterium fuses to produce tritium and protons plus energy (which may take the form of heat). In reaction (2) the products are helium–3 and 2.45 MeV neutrons, and in reaction (3) the products are helium–4 and 23.77 MeV gamma rays. There are also other branches that could be considered, but they occur with even lower probabilities. It is important to point out that deuterium could also fuse with hydrogen or tritium. The reaction chain utilized in most conventional fusion research, for instance, is D-T fusion, or the fusion of a deuteron and a triton, which produces a different set of reaction products. (D-T fusion is perhaps the easiest to produce and has the highest energy yield, making it the best candidate for commercial energy production.)

The important point here is that, no matter what the variation in the products, the relationship of the products in a particular branch remains constant. Thus a D-D reaction that produces 2.45 MeV neutrons (as was claimed in the *JEAC* paper) should produce both a quantity of helium–3 and energy that is commensurate with the number of neutrons, and the energy equivalents of these products should add up to the original mass-energy equivalent of the reactants (the two deuterons). These branching ratios have the status of experimental facts, and the most common reactions (of which D-D fusion is one) have become a routine and integral part of not only the conceptual framework but also the experimental practice of nuclear physics.[39]

For readers of the *JEAC* paper looking for fusion, Fleischmann and Pons's interpretation of their data questioned the general applicability of the conventional fusion branching ratios. Fleischmann and Pons's cold fusion thus became a disturbing anomaly for most scientists. Given that Fleischmann and Pons claimed to have measured neutrons and tritium, the most likely assumption was that reactions (1) and (2) were at work. Yet according to these ratios, a reaction that produced excess energy on the order of even 1 watt should also produce around $10^{12}$ neutrons per second. Fleischmann and Pons reported neutron fluxes of only 4,000 neutrons per second, far short of the expected value. It is for this reason that one of the early ironies of the cold fusion claims was that Fleischmann and Pons were still alive. If the experiment had produced as many neutrons as it was supposed to given the amount of heat it produced, then all the experimenters in the room (who lacked appropriate shielding) would have died instantly.

While almost all scientists recognized this discrepancy as a problem, the majority were not enthusiastic about the possibility of a radical revision to nuclear physics and quantum mechanics. For this reason most of the theoretical discussions at the end of March and into April 1989 tended to focus on trying to find reaction pathways that would account for the anomalous results. What theorists needed to find was a reaction or set of reactions with a relatively high cross section (probability) that produced a great deal of energy relative to the amount of neutrons.

One suggestion was offered by Cheves Walling and Jack Simons, two professors in the chemistry department at the University of Utah (Walling and Simons 1989). In the weeks following the press conference, Fleischmann and Pons had begun to detect He–4 in their cells using mass spectroscopy. Walling and Simons proposed the rare reaction (3) as an explanation for the lack of neutrons and tritium. Moreover, they suggested that the behavior of deuterium in the metal lattice was such that energy that would normally go to radiation went to excess heat instead (thus providing an explanation for why Fleischmann and Pons detected some radiation but not enough to account for fusion). Walling and Simons were the first to publish a possible theory of the experiment, but there was no agreement as to its plausibility; the theory simply required too many "miracles."[40]

In addition to the problem of the inconsistent fusion products, there was also the issue of whether fusion could be taking place at all in any form, especially under the experimental conditions of normal temperature and pressure that were described in the *JEAC* paper. As noted above, atomic nuclei repel one another, and for fusion to occur this energy of repulsion, known as the coulomb barrier, has to be overcome or undermined in some way. In conventional fusion this is usually accomplished through a massive application of energy, a feature that was noticeably absent in the Utah experiments. Alternatively, however, fusion can also be effected by reducing, undermining, and/or screening the coulomb barrier in some way. This is the case for muon-catalyzed fusion.[41] Steven Jones, as we have seen, developed his cold fusion experiments as an offshoot of his work on muon-catalyzed fusion.

The main idea is to take advantage of the quantum mechanical probabilities that some deuterons will fuse if held in close enough proximity.[42] The closer the deuterons get and the longer they stay there, the greater the chance of fusion. This seems to have been the basis for Fleischmann and Pons's interpretation of their results: the deuterons are held in close proximity long enough to allow fusion by a combination of electrostatic pressure and the lattice structure of the palladium metal itself. So in addition to trying to account for the heat-neutron discrepancy, theorists were also faced with the task of figuring out how the coulomb barrier could have been overcome or screened in the Utah experiments.

Several theorists performed calculations to figure out how the reduced distance of deuterons inside the palladium metal could enhance the chances of fusion, but they quickly found that the upper limits for fusion under these conditions made the possibility of fusion astronomically unlikely.[43] Such calculations served to reinforce the anomalousness of cold fusion by limiting the range of conventional explanation, and as available theoretical resources to account for coulomb screening were exhausted, more radical explanations were proposed. Letters like the following began to appear in science journals and magazines such as *Nature*: "In my theoretical investigations of the electronic structure of the $H_2$ molecule . . . I have found that the two nuclei and the electron can form a collapsing quasi-molecule. . . . it is clear that the electrons present in the matter are responsible for the Coulomb-barrier tunneling, and that the process which has been observed depends on quasi-molecular systems" (Gryzinski 1989). For many scientists the idea of proposing new particles and molecules was simply going too far, and theoretical proposals coming from all quarters seemed to be bordering on the ludicrous. Intending to satirize this situation as well as point out the severe anomalousness of the cold fusion claims, Edward Teller at one point proposed that cold fusion be explained by the existence of a new particle named the "Meshugatron."[44]

The scientists who perceived theoretical opportunities in cold fusion were extremely few in number. Most of the participants who had been drawn into the controversy—physicists, chemists, and others—viewed branching ratios as an unquestionable feature of both explanatory and experimental environments. If the reported phenomenon was D-D fusion, then the products would have to be present in the appropriate amounts. This meant that efforts at replication should be focused on finding those products that could be due only to fusion—neutrons, tritium, gamma rays, and helium. If the products were found, they would be clear evidence of fusion, and the question of how it could happen in the first place could be investigated later. If the products were not found, it was unlikely that fusion was occurring, and there was probably another explanation. What the controversy was ultimately about then was whether or not, under the conditions specified by the *JEAC* paper (which were not all that clear), the products appropriate to nuclear fusion could be measured reliably.

## Cold Fusion as a Materials Science Problem

While it seems evident from the reports and interpretations of core-set participants in 1989 that cold fusion was being primarily configured as a form of nuclear fusion, it is important to point out that other less prominent configurations were also taking shape. For a number of electrochemists and materials scientists, for example, it mattered that Fleischmann's name in particular was associated with cold fusion. Fleischmann was well known for his work on the properties of metal hydrides, and indeed the attraction for many electrochemists was simply his participation. In some cases, reputation alone may have been enough to justify researchers' participation. As one electrochemist I interviewed commented, "I was immediately somewhat skeptical and had it not been for the fact that Martin Fleischmann was the person who announced it, I think I would have dismissed it entirely. . . . Martin Fleischmann then and now actually is the preeminent experimental electrochemist . . . there is no electrochemist alive who has the credentials that Martin Fleischmann has.[45] In other cases, it was clear that researchers with similar backgrounds, who worked on similar topics, had better access to information about how to perform the CF experiments than most other would-be replicators. As John Bockris, a prominent electrochemist, writes, "A quick start was made by my own group . . . partly due to my personal knowledge of Martin Fleischmann, who readily told me on the telephone some aspects of the technique he and his collaborators had used" (Bockris 2000, 103).

For other researchers, the hook was not Fleischmann so much as their understanding of the experimental method. At the beginning of his book on cold fusion, Mizuno writes, "I was amazed to see this news. For one thing, I myself had been performing exactly the same kind of experiment for more than twenty years, but I had completely overlooked the reported phenomenon" (Mizuno 1998, 32). Mizuno goes on to describe how his Ph.D. work, which involved measuring properties of titanium hydrides, prepared him for cold fusion research and helped direct his own line of investigation. In media reports, the textual juxtaposition of Fleischmann, electrolysis, palladium, and deuterium, which may have mattered less to physicists, constituted familiar territory for electrochemists and materials scientists. Consequently, for these experts there were different cues that enabled them to draw inferences as to what was going on and how they might participate.

These different cues might be understood in disciplinary terms, but it was not a matter of physicists being skeptical of cold fusion while chemists were not. There were many scientists trained in physics who strongly believed the CF claims, and many chemists who were very skeptical. Instead, the disciplinary difference, to the extent that one can be identified, has more to do with how the problem of cold fusion was defined. Physicists were drawn to the controversy to consider the problem of nuclear fusion, and cold fusion became a problem for electrochemists and materials scientists precisely because the anomalies of excess heat and nuclear ash were appearing in a very well understood electrochemical process. In general, this led those scientists (physicists and chemists) who saw CF as a physics problem to concentrate on issues of measurement—how to detect the CF effects—whereas scientists who perceived CF as a materials-science problem tended to focus more on issues of procedure—how to produce the CF effects.

The cold fusion controversy as an instance of public science becomes important in this regard. It is likely that the distance between the disciplines of nuclear physics and electrochemistry is such that cold fusion's appearance in a specialized journal of either discipline would have significantly altered the composition of the core set of experimenters attempting to replicate the phenomenon, thereby altering the trajectory of the controversy. Thus one consequence of the public configuration of cold fusion in the media was that the phenomenon could be conceptualized as a problem in both nuclear physics and materials science simultaneously. I am not saying that the cold fusion controversy can be understood simply in terms of differences in the disciplinary perspectives of participants. As I will illustrate further in the next chapter, one could be trained in any discipline and still see cold fusion as a nuclear-physics or a materials-science problem; the battle then was over precisely what kind of problem it was.

## *The Core Set Takes Shape*

Anchored by the statements of the *JEAC* paper, the media representations of cold fusion combined with scientists' expectations to produce a dominant interpretation of Fleischmann and Pons's experimental claims as a problem in nuclear physics. One of the understandable consequences of this is that the core set came to be occupied by a significant proportion of experts in that field. And indeed, in the first few weeks of the controversy the experts most called upon to comment on cold fusion in the media were prominent physicists working at major fusion research laboratories around the world. Sound bites from physicists such as Harold Furth, the director of the Plasma Physics Laboratory at Princeton University, and Ronald Parker, the director of the Plasma Fusion Center at MIT, appeared frequently in early media reports, and while their notoriously critical quotes were used mostly to generate a heightened sense of controversy, their presence helped to reinforce the idea that the most relevant experts for the investigation of cold fusion were nuclear physicists.

As noted above, not all physicists were skeptical of the idea of cold fusion. While some viewed cold fusion as something akin to heresy, others more or less embraced the phenomenon and saw cold fusion as an opportunity to advance understanding. For physicists like the Nobel laureate Julian Schwinger and for Peter Hagelstein, a theorist working at MIT, the initial announcement of cold fusion, couched as a materials-science problem and not a physics problem, became an exciting project for developing quantum mechanical theory. They were interested in how quantum mechanics might be differently applied in the solid-state conditions of the palladium cathode rather than the pure plasma states that provide the empirical conditions for the application of most theorizing in nuclear physics.

So who precisely constituted the core set of the controversy? Some participants figured prominently in discussions of the phenomenon in the media, while others were more active in the context of scientific meetings, letters to journals, and the publication of articles. Not all core-set members performed experiments themselves, nor were they necessarily engaged in theoretical work, but some such as Richard Garwin, Harold Furth, and John Huizenga, exerted significant influences on others' interpretations of experimental data and theory. It is also worth noting that while I have outlined the core set in terms of those actors who engaged each other at a more or less public level (in the media, at conferences, in publications, or through letters and e-mail), it is actually my experience that many institutions housed their own micro core sets. This was especially true of the University of Utah, but also of research institutions like Texas A&M University, Caltech, Stanford University, and

MIT, as well as a number of national laboratories in the United States such as Los Alamos and Brookhaven. Within these institutions there were often multiple research groups across a number of departments, and this situation came with specific local modes of communication (colloquia, internal memos, and e-mail) as well as institutional politics. Finally, as will become evident as my story develops in later chapters, the initial core set as I have begun to describe it here is heavily biased toward the experience of the controversy in North America and to some extent Europe. While the dynamics of the controversy in the United States certainly had a profound if not overriding effect on cold fusion research in the rest of the world, it is important not to overlook the differences in the development of core sets in other countries, especially Japan, Russia, Italy, and France.

## *Of Audiences and Objects*

The discussion so far has been meant to highlight the dominant reading of media reports and scientific literature at the very beginning of the cold fusion controversy. While I have analyzed only a small portion of the media coverage of the early weeks of the controversy and presented a small segment of the technical issues raised by the *JEAC* paper, my goal was to convey the way in which media reporting and the paper worked together with audiences of scientists to produce a core set of researchers who predominantly viewed cold fusion as a problem for nuclear physics.

It would be an exaggeration to claim that media representations, no matter how credible, were the sole cause of replication attempts or even of the theoretically based skepticism or opportunism of scientists. Rather, my argument has been that such representations became kinds of conceptual resources for making decisions in the context of reading the *JEAC* paper and trying to reproduce the experiment or propose a theoretical explanation. The texts of the media reports by themselves are incomplete; they do not present the whole picture. But the text of the *JEAC* paper is not necessarily any more helpful. The texts lack detail and clarity, and this lack was filled by the expertise and background knowledge of the readers. In this sense, the readers completed the text, and the text worked to invite the readers to complete it.[46]

The crucial interaction then, at least in the first few weeks of the controversy, was not between Fleischmann and Pons and would-be core-set members, but between would-be core-set members and these media texts. The object or the experimental entity understood as "cold fusion" was produced in the interaction between the text and its expert readers.

There are two important consequences of accounting for the beginning of the controversy in this way. First is the basic constructivist point that a different combination of texts and readers would probably have produced a different conception of the entity under investigation and a different outcome for the controversy about its existence. While there is no definitive way to prove such a hypothetical claim, the idea is to establish the conditions for explaining in sociological terms how multiple and divergent understandings of the same phenomenon are possible without assuming that one or another understanding is a priori right or wrong. Traditionally much of the research in science and technology studies has been aimed at showing how objects of science that we take to be real or true (such as Newton's laws or the existence of electrons) are constructed.[47] What I want to begin to argue here is that objects of science that we do not take to be real or true are also constructed and that such objects, classed as failures, errors, fantasies, or frauds, circulate in and amongst those other objects we have now come to count on in our daily lives.

My second point concerns the question of who ultimately is responsible for the configuration of cold fusion. It might seem that the most likely sociological question to ask here is, Why did Fleischmann and Pons opt to present such a controversial claim? We know they did not need to couch their claim in terms of nuclear fusion, and indeed their experimental results could have been presented in any number of ways. As a result, we might be led to suspect that they opted for a fusion hypothesis knowing that the risk was great but that the potential reward was worth it if they were right (Pinch 1985). Or perhaps they simply made a strategic error and misgauged the open-mindedness of their audience. Such questions may be interesting, but they limit us to providing an account of the behavior of Fleischmann and Pons and detract from the task of producing a genuine sociology of knowledge about cold fusion. In concentrating on the public representation and reception of cold fusion, I want to de-emphasize the notion of authorship entailed by looking at Fleischmann and Pons as claims-makers (either in the lab or in texts). Once deployed in public contexts, cold fusion takes on its own life distinct from its material and textual origins. Fleischmann and Pons may resist or deploy this public cold fusion, but the object ultimately escapes their control.

The consequences of this situation as I analyze it in the rest of this book are twofold. First, the configuration of cold fusion as "fusion" brings unanticipated groups of actors to the core set. As we shall see in chapters 3 and 4, the participation of these actors, backed by powerful networks, becomes an important element in the closure of the controversy. Without the public representation of cold fusion, without the participation of these actors, the controversy could have ended differently. Second, the configuration of cold fusion as "revolutionary" brings other unanticipated actors to the periphery of the core set. These less powerful actors tend to occupy nonstatus roles in the world of mainstream science, and their actions have little direct effect on the outcome of the controversy. Yet as I argue in chapter 6, after closure, these peripheral actors become crucial to cold fusion's continued survival. In this way, the public representation of cold fusion becomes implicated in both the death of cold fusion and its afterlife.

# Chapter 3 The Cold Fusion Controversy

## *High-Temperature Superconductivity*

In January 1986, Alex Müller and Georg Bednorz announced the discovery of "high-temperature" superconducting materials, for which they won the Nobel Prize in 1987. Like the announcement of cold fusion, the announcement of high-temperature superconductivity (known as $HT_c$) came as a surprise to scientists. Previously, certain materials could become superconductors only at extremely low temperatures, and, with important commercial applications at stake, scientists working in the early 1970s had tried and failed to produce materials that would be superconducting above 23 degrees Kelvin.

As with the case of cold fusion, Müller and Bednorz's claim was considered controversial (Felt and Nowotny 1992). In the first instance, other scientists perceived Müller and Bednorz as being outside the mainstream of superconductor research, and their work had been performed at a small IBM research laboratory in Zurich that had no reputation in the field. In the second instance, their claims contradicted conventional beliefs and a constellation of practices associated with the study of superconducting materials.[1] Yet from the beginning, as with cold fusion, newspapers latched on to the story as a triumph of "little" science, and the media presented the production of $HT_c$ materials as a simple process that could be done in any high school laboratory with the proper equipment.

Interestingly, however, it could be argued that greater skepticism was initially expressed amongst scientists over Müller and Bednorz's announcement than over Fleischmann and Pons's. The reason for this stems from the prior existence of a superconductor research community that had already attempted and then ruled out the possibility of $HT_c$. Because others had already tried to exceed the 23-K cap, a climate of skepticism existed before Müller and Bednorz's claims were published (and even before they did their research). In Fleischmann and Pons's case, the idea of cold fusion may have seemed unlikely given the assumptions of basic nuclear physics, but because the claim was so novel scientists may have been more willing to entertain the possibility. Despite the initial skepticism in the $HT_c$ community, however, within weeks of the announcement, Müller and Bednorz's material was reproduced by nearly every research group that made the attempt. In addition, several of the replicating groups were able to innovate on the original material to produce even better superconductors.[2]

At the time of Fleischmann and Pons's announcement, the memory of the excitement of the race for higher-temperature superconductors was still fresh in the minds of both scientists and science reporters. Perhaps Fleischmann and Pons were another Müller and Bednorz? Perhaps their claim of 400% excess heat was just the tip of the iceberg? Perhaps, like high-temperature superconductivity, there may have been no generally accepted explanation for the phenomenon, but the experimental effects were nonetheless real. In this sense, cold fusion represented an intellectual and economic possibility that rivaled and even exceeded its precedent in the discovery of $HT_c$.

Yet it quickly became evident that there was a crucial difference between cold fusion and $HT_c$. Müller and Bednorz's claims were immediately reproducible and Fleischmann and Pons's claims were not, and in the end this is crucial in accounting for the different fates of the two phenomena. The story of $HT_c$ is interesting here, however, not because it illustrates how high-temperature superconductivity is real and cold fusion is not. Nor is the story useful for illustrating the difference between "good" and "bad" science. What makes the story of $HT_c$ different when viewed from the perspective of the sociology of scientific knowledge is that right from the outset, starting with Müller and Bednorz's original claim, all the participants had a clear idea of what they were looking for—they knew what superconductivity was. Amongst the research groups that worked on superconductors, there was some difference of opinion about how superconductivity worked and what might be the best theoretical explanation, but there was no disagreement over how one recognized or measured whether a given material was in fact a superconductor. As a consequence, the issue of whether any given attempt at replication was or was not successful was not a subject of controversy. In the case of cold fusion, the problem of replication was at the heart of the controversy.

## Cold Fusion and the Experimenter's Regress

As with the discovery of $HT_c$, the unexpected and unconventional nature of the cold fusion claim combined with social, economic, and intellectual implications to make the issue of experimental replication a priority for researchers. Fleischmann and Pons claimed that they had reproduced the anomalous effects a number of times, and acknowledging that something unusual and unlikely was occurring, they appealed to independent replication for confirmation.[3] Thus the idea of experimental replication was conventionally held by all participants to be the court of appeal in the controversy. Fleischmann and Pons acknowledged that they would be "right" only if others could repeat their experiments, and they would be "wrong" if their experiments could not be repeated. This statement seems reasonable in principle, and it speaks to the near universal recognition of the importance of replicability in the adjudication of experimental claims, but it betrays a range of difficult and problematic issues concerning how one recognizes when an experiment has been repeated successfully or not.

In the case of $HT_c$, a material is superconducting if it has a critical temperature at which electrical resistance drops to zero, if its critical temperature drops to a lower value in a magnetic field, and if it exhibits the "Meissner effect."[4] These criteria for the recognition of superconductivity had been established prior to Müller and Bednorz's discovery. Müller and Bednorz did not dispute the criteria; they only disputed the accepted limit on the critical temperature. Superconductivity was already a "black box" (Latour 1987) for scientists. As a consequence, it was a priori possible for Müller and Bednorz, and their would-be replicators, to create a "false" high-temperature superconductor—a material that they thought might be superconducting but turned out not to be because it failed to exhibit the Meissner effect, for instance. The criteria for establishing superconductivity were simply never in doubt.[5]

This was not the case for cold fusion. Conventions for recognizing (detecting or measuring) fusion reactions already existed prior to Fleischmann and Pons's announcement, but the effects reported by Fleischmann and Pons did not seem to meet any of the established criteria. Rather than accepting that they had measured a "false" fusion signal, Fleischmann and Pons argued that they had in fact detected a new kind of nuclear process that did not meet any of the established criteria for recognizing fusion. As a consequence there were no prior *communally* accepted definitions for how this new kind of fusion could be recognized. There was no prior Meissner effect test for cold fusion. This does not mean that nobody knew what a test for cold fusion would be. Most nuclear physicists involved in the early days of the controversy, for instance, had a pretty good idea of what they should be looking for. (Some claimed that the detection of 2.45 MeV neutrons was a fairly definitive test.) The problem was that not everyone who was involved shared similar ideas about what to look for, and this meant that the question of how to test for cold fusion became the subject of disagreement.

At the heart of the controversy over the existence of cold fusion was a situation that has been characterized in the sociology of science as "the experimenter's regress." According to Collins, "The Experimenter's Regress arises when an experiment is done without knowing what the right result should be. That is, the Regress applies in those important cases where an experiment, or series of experiments, is done in order to establish certainty. It occurs when there is dispute at the outset. In such cases it is not possible to inspect the outcome to know whether or not the experiment has been done properly for the experiment is meant to establish what the outcome is" (Collins 1989, 87–88). The key feature of the regress in the case of cold fusion is that at the outset of the controversy there was no collectively acceptable definition or characterization of what cold fusion was. Without a generally accepted characterization of the phenomenon in question, would-be replicators simply cannot have any epistemic guarantee that their experiments are a competent test of the existence of the phenomenon. More simply put, how does one know one is "seeing" cold fusion when no one is sure what cold fusion is? Without knowing beforehand whether cold fusion exists or not, one might view any potential confirmation as a result of experimental error, and any disconfirmation as missing an important element of the proper experimental protocol. Yet everyone involved agrees that to decide whether cold fusion exists, Fleischmann and Pons's results have to be replicated. The closure of this kind of controversy thus depends on the participants' ability to resolve the regress by establishing the equivalent of a Meissner effect test for cold fusion.

The implicit assumption of the argument in this chapter is that the establishment of experimental criteria for recognizing cold fusion is a constitutively social accomplishment and is no less scientific, objective, or rational for being so. Numerous case studies of controversy trace various kinds of social dynamics whereby core-set participants arrive at "agreements" as to what will count as a successful and unsuccessful replication of a controversial phenomenon. Many of these case studies have been presented with an eye toward proving what I intend to take for granted. The notion of closure I propose assumes that actors are predisposed to the resolution of conflict in order to enable or sustain different levels of collective action aimed at the production of knowledge. Closure can then be viewed as a consequence of participants' social and material maneuvers to establish the existence or nonexistence of the controversial phenomenon.[6]

The aim of this chapter is to develop the concept of closure as a collective social and material gradient of practice that actively resists dissension. This is an idea that is closely allied with actor-network approaches in science studies, and it differs importantly from neo-Kuhnian approaches to understanding closure as collective action in terms of consensus or near consensus of belief or practice (Kuhn 1970; Collins 1985). In the case of cold fusion, I argue that collective "agreement" on the facts of the matter comes about not only because participants recognize the rational force of certain arguments over others or because certain perspectives better suit their "interests," but because an unequal distribution of power develops, which enables some actors and disables others. Collective agreement is secured ultimately through changes in the composition of the collective, both through the "enrollment" or addition of social and material actors (Callon 1986; Latour 1986) and through their exclusion or negation. In chapter 7 I will develop this idea of closure in more theoretical detail, but what follows here is an account of the development and closure of the controversy as it revolves around the issue of experimental replication.

## *Replication Reports as Allies*

Since the end of the controversy in 1990, we have been able to look back and point to reproducibility as the key difference between the case of high-temperature superconductors and the case of cold fusion. Almost every group that tried to replicate $HT_c$ was successful, while the overwhelming majority of groups that tried to replicate cold fusion in 1989 were not. In the first few weeks after the March 23 announcement, however, this did not actually appear to be the case. In fact, Fleischmann and Pons's claim, like Müller and Bednorz's discovery, seemed to meet with immediate success as one research group after another reported the measurement of one or more of the anomalous effects associated with cold fusion. On April 1, 1989, researchers at Tokyo University claimed to be observing excess heat and gamma rays in their CF experiments. On April 2, another report, this time the observation of neutrons, came from researchers at Kossuth Lajos University in Hungary. On April 6, Kevin Lynn, a physicist at the Brookhaven National Laboratory, announced that his group had detected neutrons. On April 10, one of five CF research teams at Texas A&M reported measuring excess heat, and another group working at Georgia Tech reported high fluxes of neutrons. In the media, this news from reputable researchers heightened the expectation that the CF effect was real, and throughout the month of April positive results came from researchers at the University of Washington (V. L. Edden and W. Liu), Stanford University (R. Huggins), Case Western Reserve University (M. Landau), Moscow University (R. Kuzmin), the Indira Gandhi Center for Atomic Research in India (C. Mathews), and the Frascati Laboratory in Italy (F. Scaramuzzi).

This flood of positive reports reinforced the enthusiastic reception of cold fusion, which peaked during a special session of the American Chemistry Society (ACS) on April 12. At the ACS meeting, Pons announced triumphantly to a crowd of seven thousand that the anomalous effects had been confirmed in several labs, and work was under way to better characterize the phenomenon. In effect, Pons was drawing on reports of successful replication (some from the media, some from word of mouth) as allies to bolster the certainty of his experimental claims in the face of skepticism from prominent nuclear physicists. Cold fusion, such as it was, was no longer a product of Fleischmann and Pons's claims but was now intertwined with the claims of other scientific groups. If one wants to understand closure simply in terms of who has the greatest number of allies, then on April 12, Pons was winning, and the enthusiasm and cheers from the scientists at the ACS meeting seemed to indicate that this was so.

Not all scientists were convinced, however, and they had ways of detaching Fleischmann and Pons's claims from the reported replications. Many, for instance, refused to accept informal reports in the media as instances of confirmation. What were successful replications for Pons were mere rumors for others: "There were rumors everywhere . . . some rumor that Bell Labs did it, a rumor that Bell Labs didn't do it. Rumor was that we had done it, and wouldn't tell anyone. People would call us up and say, 'We know you've done it! We know you're lying about it too!' We heard Florida had seen tritium. Ten minutes later I was on the phone with the guys in Florida to figure out what they did. I was on the phone 12, 15 hours a day over three time zones."[7] Other scientists questioned reports of replications that they did not recognize as coming from credible experts in their field: "It struck me as odd that the positive replications came from people I had never heard of in the electrochemical field whereas the people who I knew and trusted were either skeptical or negative. Why is it that only in Hungary you could replicate this thing, but at Stanford you couldn't?"[8] Evaluations of this kind stress the ways in which positive reports could be discredited (and hence no longer be enrolled as allies) based on the context of their appearance (i.e., in the media, or outside researchers' normal disciplinary networks).

A different and more important issue was that the content of replication reports was not necessarily in accordance with the claims of the original experiment. As the initial confirmations became allies for the claims of Pons and Fleischmann, so the nature of the cold fusion claims began to change. Some groups reported excess heat but detected no neutrons or tritium; others reported tritium but no heat. Some groups reported results using titanium electrodes rather than palladium; other groups used a different electrolyte or electrode preparation. Could all these reports be counted as successful replications of the same phenomenon? Such problems were compounded by the fact that many of the researchers who had reported positive results were either unsure of the precise protocol by which they obtained their results and thus were having difficulty repeating their own experiments or else were barred from discussing the details because of patent applications that had been filed by their institutions. There was then no clear way to interpret the positive results. For the moment at least it seemed that the detection of any anomaly in experiments loosely structured around the idea of electrolyzing metal in the presence of deuterium could be considered a confirmation of cold fusion.

## *Turning Positives into Negatives*

Some critics of cold fusion referred to the wide variation in the reported confirmations (as well as the method of the reporting, i.e., press conference as opposed to peer review) as evidence for the claim that the cold fusion effects could probably be explained as artifacts. Support for this view came quickly as two of the most prominent positive reports (at least as far as the American scientists and media were concerned) were retracted. A finding of high levels of neutrons at Georgia Tech turned out to be due to the temperature sensitivity of $BF_3$ neutron counters. James Mahaffey appeared on television holding the probe in his hand—his body heat generated a neutron flux! Mahaffey publicly apologized, retracted his claims, and stopped working on cold fusion. The episode prompted critics to question the neutron results of other groups, including Fleischmann and Pons, who had also used a $BF_3$ counter (although not the same model as the one used by Mahaffey). Yet on its own, this retraction was not enough to unsettle the cold fusion claims; others, like Steven Jones, had detected anomalous neutrons without $BF_3$ counters, and there were also other measures of nuclear ash: helium, tritium, and gamma rays.

A second retraction came from Charles Martin, an electrochemist at Texas A&M and a friend of Stanley Pons. In the first week of April, Martin's group detected the production of nearly 90 percent excess energy in some of their cells. Problems started arising, however, when Martin ran a series of control experiments using normal water and a blank carbon electrode. To his surprise, the control experiments produced excess heat. Two weeks later, after further investigation, Martin retracted his results and withdrew a paper he had submitted to a journal.[9] The group had determined that its thermistor had not been properly grounded, thereby providing an unaccounted-for source of heat. When the error was corrected, most of the excess heat disappeared.[10]

The experiences of both Mahaffey and Martin are notable in that an "apparent" anomaly disappeared once certain adjustments in measurement were made. This helped set conditions for what counted as the detection of a real anomaly. Thus in the presence of a $BF_3$ counter, neutron results were suspect. In the presence of an ungrounded thermistor, excess heat results were suspect. As a result of this, critics became concerned that reports confirming cold fusion effects should provide more details about the instrumentation used in the experiments, and those reports that presented few details began to be viewed with suspicion. But while the early retractions of Mahaffey and Martin certainly raised suspicions, they were hardly damning evidence against cold fusion. In other confirmatory experiments, neutrons were appearing in the absence of $BF_3$ counters and excess heat was appearing in the absence of ungrounded thermistors.

## *Huggins, Furth, and the Light-Water Control*

If a retraction wouldn't unsettle cold fusion, perhaps a more definitive control experiment could. Harold Furth, the director of the Princeton Plasma Physics Lab (PPPL), was, as we have seen, an early critic of Fleischmann and Pons's claims. For Furth, the claims simply made no sense in terms of nuclear physics, and he proposed a crucial test in the form of a control experiment. Furth suggested using normal "light" water ($H_2O$) instead of heavy water ($D_2O$) in an effort to control for the anomalous effects. The idea was that if D-D fusion was occurring, the measured effects (heat, tritium, and neutron production) should vary in proportion to the amount of deuterium present.[11]

On April 26, 1989, Furth, Fleischmann and Pons, Jones, and several other CF researchers were invited to testify before the U.S. House Committee on Space, Science, and Technology. The committee had convened to review the cold fusion claims and to make recommendations with respect to federal funding for future research. In his statement before the committee, Furth made a version of the light-water test a control not only for the existence of the phenomenon, but also for funding: "The decisive control experiment would be to mix small quantities of heavy water into the light water: the 'excess heat' from the H-D reactions should be observed to rise proportionally as the fraction of heavy water is increased. . . . a systematic finding of large 'excess heat' in heavy-water experiments and 'no excess heat' in the light-water control experiments would provide a significantly encouraging sign in favor of the 'cold fusion' hypothesis."[12]

The issue of a light-water control had become significant once scientists got a copy of the *JEAC* paper and saw that there was no mention of control experiments. Fleischmann and Pons were put on the spot, and their reply was troubling. Pons in particular was evasive on the question of the control, saying, "A baseline reaction run with water is not necessarily a good baseline reaction. . . . we do not get the expected baseline experiment. . . . we do not get the total blank experiment we expected."[13] Had Fleischmann and Pons observed excess heat in light water, as Charles Martin had? If so, it was likely that the excess heat was coming from some chemical reaction or was the product of an instrumental artifact. Why weren't Fleischmann and Pons explaining how they did their experiments? Why were they being so cryptic? Some suggested that their hands were tied because of patent restrictions; others suggested that they did not have enough data to talk about their experiments competently. Four years later, at the third International Conference on Cold Fusion (ICCF–3) in Japan, some CF researchers would indeed argue that nuclear reactions were taking place in light-water experiments, making cold fusion even more of an anomaly, but this was something Fleischmann and Pons were not ready or willing to entertain as a possibility (at least publicly).[14]

As with the cases of the Georgia Tech and Texas A&M retractions, however, cold fusion was saved by other replication reports. Robert Huggins, a senior materials scientist and the founder-director of the Center for Materials Science at Stanford University, held a press conference on April 18 announcing that his group had measured excess heat in some cells (although not as much as Fleischmann and Pons had measured). More importantly, Huggins also ran a series of control experiments using light instead of heavy water. The light-water cells produced no discernible excess heat, and Huggins presented this as evidence that the reaction might be nuclear in origin.

Huggins's argument suggested that his experiment, and cold fusion along with it, had passed Furth's test. In his own testimony at the House Committee hearing, he argued that, "as was pointed out . . . the validity of the reported results would be greatly enhanced if there were direct experimental evidence of a significant difference in behavior between the hydrogen–light water–palladium system and the deuteron–heavy water–palladium system. Such experiments would subtract out any contributions from spurious chemical effects, for they would be present in both. We . . . have undertaken experiments that addressed that question. . . . the results that we have obtained lend credence to the Fleischmann and Pons contention."[15]

The addition of Huggins's experimental results, along with those of other positive replications, supported cold fusion through the extension of a network of allies (researchers, reports, laboratories, and experiments) that together could resist criticisms that might discredit any one on its own. While it was simple enough to be suspicious of Fleischmann and Pons's results alone, it was more difficult when these results were coupled with those of Steven Jones or Robert Huggins, amongst others. For critics, the prospect of dealing with each and every confirmation was turning into a monumental task, and in the terms of actor-network theory, cold fusion would be able to survive as long as the network of associations that made it up could resist attempts by critics to dissociate it.

In the discussion so far, I have focused on accounts of experiments that appeared in public contexts (journal articles, newspapers, conferences, government hearings). It is in these contexts that associations are made or broken.[16] Up until the end of April, the association of allies in support of cold fusion had managed to withstand the dissent of individual and even institutionally powerful critics like Furth. Everything was going well for cold fusion. Fleischmann and Pons were initiating a new set of experiments, theorists were hard at work on possible explanations, and new confirmations were appearing every week around the world. The U.S. Department of Energy directed its national laboratories to begin serious investigation, the governor of the state of Utah signed a bill appropriating seed funding for a cold fusion research institute, and Fortune 500 companies, along with the news media and the general public, remained intensely interested in the phenomenon.[17] What changed the course of events was the public formation of a counter association of allied actors that would become "stronger" than cold fusion in part by achieving a superior strength in numbers (numbers of scientists, publications, laboratories, and experiments), but also by eliminating the capacity of its opposition to act.

## *The Decline of Interpretive Charity*

On May 2, 1989, at a special double session of the annual meeting of the American Physical Society (APS) in Baltimore, cold fusion suffered a major blow. The APS meeting became the first occasion on which researchers who had not measured anomalous effects in their CF experiments presented their data. Previously, only positive reports had been given any public attention. The sessions were attended by around a thousand scientists, and of the forty abstracts submitted and nineteen papers presented, none came out in support of cold fusion. More importantly, the meeting marked a point in the controversy when some would-be replicators began to claim that their inability to observe cold fusion effects was evidence that cold fusion was not real. What had appeared to be merely reports of unsuccessful confirmations could, after the meeting, be viewed as successful negative replications.

Up until the APS meeting, scientists who had been unable to detect CF effects might assume that they had been missing some crucial element of the experimental protocol. This seemed a reasonable assumption given the lack of detail in the *JEAC* paper, and at the same time, would-be replicators were aware that Fleischmann and Pons had worked on their experiments for over five years and that they were experts in the field of electrochemistry. It made sense that scientists unskilled in electrochemistry working on cold fusion for only a few weeks might not be able to perform the experiment properly. In addition, other reputable scientists had purportedly been successful in reproducing some of the CF effects. Finally, even if Fleischmann and Pons's interpretation of their data in terms of cold fusion was wrong, their skill and reputation suggested that they certainly could have discovered a genuine and significant energy-producing anomaly.

This interpretation was even more plausible in light of the rumor that Fleischmann and Pons were being secretive about the complete protocol for their experiments. Whether they were justified in this or not, it seemed likely that many scientists were working with less information than might be required to successfully replicate the effect. Indeed, a precedent existed for this in Paul Chu's discovery of a 90-K superconductor; an error in the published composition of the material had the effect of holding back successful replication for a number of days. Many believe the error was intentionally published to give Chu's group a head start in further innovation (Hazen 1988, 90–96). Perhaps Fleischmann and Pons were behaving in a similar fashion?

As a consequence of all these factors, many experimenters were (and still are) prepared to give Fleischmann and Pons the benefit of the doubt. One physicist explains it this way:

> Today we specialize. One scientist may be a hydride specialist, another a fusion specialist, and somebody else might be an electrochemist, but few people have a complete understanding of all these disciplines. When cold fusion was reported, the tendency was to think about it in terms of what you knew in your own area. Although it didn't look plausible, many thought that there might be something in another area that they didn't know about that was allowing fusion to occur. Since no respected experts stood up and said that the conclusion

> about fusion was clearly erroneous, giving specific reasons, everybody was willing to give it the benefit of the doubt. The thinking was that maybe the fuzz factor was in some area outside your expertise. (Lawrence Livermore Nuclear Laboratory 1990)

A materials engineer I spoke with compared the lack of information from Fleischmann and Pons to the clandestine work of Alex Müller and justified the "benefit of the doubt" in these terms:

> At that time I also got involved in high-temperature superconductivity and my argument was—well Müller the co-discoverer told us that—I asked him how did you even dare to do such an experiment—you know insulating oxide to measure superconductivity, and he said, in great secrecy because he didn't want to be ridiculed by his peers, and yet he did because he had this tremendous insight—crystal structure and conductivity and it worked out he got the Nobel Prize. And I said well maybe this is the same thing, everybody is skeptical, but who knows maybe it's something we never dreamed of, and that's the history of science you know. . . . so I feel that a scientist should have an open mind and go by the evidence rather than what the textbooks say.

The degree to which scientists were willing to extend Fleischmann and Pons (and cold fusion) the benefit of the doubt is reminiscent of Collins's (1981c, 53) use of Gellner's notion of "interpretive charitability" (Gellner 1970). Collins's discussion of the concept is brief, and perhaps there is a need to respecify its use here. The idea of interpretive charitability is useful for referring to actors' interpretations of each other's actions in the light of their perceived experience. Consequently, I want to specify "interpretive charitability" to mean the degree to which one will doubt oneself before doubting another in the face of seemingly contradictory experience. Thus experimenters who find their experience in contradiction with that of Fleischmann and Pons are exhibiting a high degree of interpretive charity if they doubt their own experience. This does not necessitate the belief that Fleischmann and Pons are right, only that they might not be wrong. Armed with this definition, it may be possible to trace individual scientists' "interpretive trajectories" (Rudwick 1985, 411) with respect to claims for the existence of cold fusion and to pinpoint if and when their interpretations become more or less charitable.

One example will help to illustrate this. Nathan Lewis led an interdisciplinary team of researchers trying to replicate the CF effects at Caltech during April and May 1989. The experiments began the day after the March 23 press conference, and like most other groups, Lewis and his team began by building a crude electrochemical cell using a combination of scrounged and store-bought materials. Lewis is an electrochemist, but he had access to more sensitive neutron detectors than Fleischmann and Pons did, so he reasoned that he would be able to achieve more precise radiation measurements, along with measurements of excess heat.[18]

The Caltech group did not detect any anomalous heat or radiation within the first two weeks, but Lewis did not become skeptical. He saw the publication of Jones's paper in *Nature* as a tentative confirmation of the CF effect, and, in an effort to improve his experiments, he sent the following e-mail message to Pons: "I am sure that you have been inundated with requests, but we have an extremely sensitive neutron counter and can confirm the reaction, however our initial attempts . . . did not yield any neutrons. Could you please send me a preprint and/or bitnet explicit directions of electrode preparation, D electrolytes, voltages etc., so that we can confirm the experiments and quantitate the neutron yield?"[19]

Lewis did not receive a response to his query, but the group persisted. With the support of the Caltech administration, Lewis launched a larger-scale attempt by running over fifty cells with slightly varied parameters (gold anodes, titanium electrodes, different electrolytes and electrode preparations, etc.). Again the group did not detect anything unusual, and Lewis sent another message to Pons: "Obviously, we are missing a key step somewhere. . . . We just haven't seen any of the effects that you have reported. We will work like crazy trying to improve our mistakes if you can offer us any advice for changing the experimental procedure. . . . thanks for your help, and I hope that we can be of help to answer the doubters regarding the validity of these observations."[20]

Again Lewis received no reply to his message, and as the date for the APS meeting approached, Caltech issued a press release in which Lewis is quoted as saying, "We have seen no evidence whatsoever for nuclear reactions or even unusual chemical reactions. . . . In the course of our calorimetry measurements, we've uncovered a number of problems. These problems may lead to errors large enough to cast serious doubts on published determinations of excess heat. When these errors are avoided, we obtain no evidence for excess heat production" (Finn 1989).

Sometime during the month of April, Lewis came to represent his attempts at replication differently. At the beginning, Lewis explained the lack of a positive result in terms of the differences between the Caltech and Utah protocols; it seemed reasonable to assume that he was missing something. After running a large number of experiments attempting to "fix" his error, Lewis's position shifted. The errors were not his; they were Fleischmann and Pons's. At the APS meeting, Lewis's position shifted even farther, and he attacked

Fleischmann and Pons publicly, saying, "One of the main things we've learned during the course of experiments is just how easy it is to fool oneself into thinking that there is an effect when there actually is none."[21]

Lewis's comments suggest that a major shift in the interpretation of his CF experiments occurred during the last two weeks of April.[22] What is important is not the precise reasons for his change of opinion, but the fact that it was possible to represent the same set of null results differently. Nothing about the data themselves had changed in those two weeks; if Lewis so chose, he could have remained charitable and either waited for a response from Pons or else continued with his investigation (perhaps asking the advice of other "successful" replicators).

On the whole, Lewis's presentation was warmly received (especially in the popular and scientific news media), and the Caltech results have been held up as one of the most important disconfirmations of Fleischmann and Pons's claims. But scientists who had reported positive results and who were not at the APS meeting were not so impressed. One electrochemist I interviewed, for instance, had the following to say about Lewis and other critics who presented results at the APS meeting: "People became quite skeptical quickly when they couldn't do it themselves immediately, which is an interestingly arrogant position. . . . if anybody were to compare one or two weeks of their own skill against three years of Fleischmann and Pons and say that these things are equal, then that is an interestingly arrogant position that I certainly wouldn't be able to take."[23]

To turn a set of negative results into a negative replication that disproves the original claim requires a judgment, a decision on the part of the would-be replicator to represent experiments differently. There was nothing about the accumulation of negative or null results alone that could disprove cold fusion; it required the conversion of null results into negative replications by individual experimenters, as well as the acceptance of this conversion by participants in the controversy. In order for this to occur and for the experimenter's regress to be resolved, the extension of interpretive charitability has to appear to participants to be the wrong choice.

## *Justifying Negative Replication*

Lewis was just one experimenter, but his became an important voice in the context of the APS meeting. As Close (1991, 214) remarks, "In one evening, primarily due to Lewis's talk and Koonin's outspoken criticism of the two chemists, suddenly the emphasis had swung the other way. Now it was the skeptics who were in the leading role." In this sense, Lewis's role (along with other vocal critics like Steven Koonin) was similar to that of the scientist Quest in Collins's account of the closure of the controversy over Joseph Weber's discovery of high-flux gravity waves: "The series of experiments [by Quest] legitimized the openly publishable statement of strong and confident disagreement with Weber's results, but this confidence came only after what one might call a 'critical mass' of experimental reports had built up, and that this mass was 'triggered' by scientist Quest" (Collins 1981c, 44). If Lewis, like Quest, was a "trigger," it remains to note what the "critical mass" amounted to. I have argued that the cumulative mass of negative experimental reports alone was not enough to dissociate cold fusion. What was needed was the mutual interaction of scientists at the APS meeting in support of less charitable interpretations of the evidence for cold fusion. The APS meeting produced not a definitive refutation of the cold fusion claims but an atmosphere of skeptical solidarity.

In the context of the APS meeting uncharitable interpretations of the experimental data supporting cold fusion were thus made publicly acceptable. Lewis and a number of other scientists used the meeting not just to present their results as unsuccessful replications; they presented their results as disconfirmations, and proceeded to offer alternative explanations for the CF effects. They launched a counterclaim. The excess heat and radiation observed by Fleischmann and Pons were due not to cold fusion, but to instrumental artifacts. The public acceptability of this claim was facilitated by three important factors: the existence of prior social networks, appeals to standards of experimental competence, and the rhetoric of similarity.[24]

### PRIOR SOCIAL NETWORKS

The deluge of criticism at the APS meeting was not a spontaneous affair. Prior to the meeting, many fusion research scientists had been exchanging information and opinions through existing channels of informal communication. Thus fusion researchers and nuclear physicists who normally communicated with one another on a regular basis often added discussions of cold fusion to other topics. Examples of this include the initiation of a cold fusion discussion thread on the private electronic network of the IBM labs, the exchange of experimental data amongst researchers working at the various DOE labs, and an electronic newsletter started by Douglas Morrison, a retired physicist working at CERN (European Organization for Nuclear Research).

Outside the gaze of the public and the media, scientists in these networks seemed to be developing a skeptical position with respect to cold fusion. Note the following exchange of e-mail between a senior scientist at an IBM laboratory and a colleague at one of the DOE labs; the DOE scientist writes, "I just received a copy of the paper by Guinan, Chapline and Moir at LLNL. It looks very good to me. Odds against Jones are now 1:1, down from 5:1. Odds on Fleischmann/Pons remain at 25:1. Is it true that Pons claims to have seen the same effect now for ordinary hydrogen?? Maybe that's why I have so few takers."[25]

After attending one of the first meetings on cold fusion held in Erice, Italy, on April 12, the senior scientist writes back, "Hope this gets to you. I have seen the Chapline paper on muon-catalyzed fusion in PdD [palladium deuteride]. I don't believe it at all. . . . Two lousy experiments. M.F. [Martin Fleischmann] also responded to me that they had obtained special hw [heavy water] with only 40 dpm/ml of T decay. I have now calculated, though, that if they find 100 dpm/ml, that amounts to at least 500 times as much as would accompany their 4000/s neutrons. Another fishy result."[26] Here, scientists are seen exchanging reports and evaluations, laying bets and making judgments. To the extent that individual evaluations are similar, the exchange allows for mutual support and recognition that the evaluation is correct.

In April and into May, the replication attempts at most of the major fusion research laboratories in the United States and Europe had not produced any anomalous results. Would-be replicators in these labs used existing communications networks to exchange ideas and information, and the various groups were able to see right away that they were not alone in getting negative results. In the absence of disagreement from "successful" replicators, there was a marked decline in interpretive charity amongst researchers who were a part of these networks.

In the context of the cold fusion controversy, network members included researchers at the Plasma Fusion Center at MIT, the Plasma Physics Lab at Princeton University, several of the DOE laboratories, the IBM Research Laboratory in Yorktown Heights, New York, Garching in Germany, Culham in England, and CERN. Supporters of cold fusion would later refer to the actions of prominent fusion scientists at these institutions as part of a conspiracy on the part of the "fusion fraternity" to suppress CF research. Whether consciously planned or not, the prior informal communication that filtered through the network supported the plausibility of the disconfirming experimental evidence presented at the APS meeting. The meeting was not so much an occasion for convincing scientists that Fleischmann and Pons were wrong as it was an occasion for a public display of collective skepticism, a forum for the public sanctioning of unsuccessful experiments (i.e., experiments producing null results) as negative replications.

## STANDARDS OF EXPERIMENTAL COMPETENCE

In addition to providing an opportunity for the public realization of a collective skepticism born out of prior social networks, the APS meeting also enabled the related development of a limited consensus around a set of conditions for what could count as a competent and credible cold fusion experiment. A number of theoretical arguments that had been made in early April were made again, only this time they were expressed as necessary experimental conditions for the replication of cold fusion. These conditions were premised on the reasonable assumption that cold fusion could be understood in terms of conventional nuclear processes. As I argued in chapter 2, this assumption became reasonable to a large extent because of the way the phenomenon was configured in the media and in the *JEAC* paper in the first weeks of the controversy.

As a consequence, it made sense to many scientists at the APS meeting that a successful positive replication of cold fusion required the measurement of nuclear products commensurate with excess heat. Experiments that measured either heat or nuclear ash but not both could not therefore count as positive replications. In addition, a competent cold fusion experiment should be able to account for the entire energy balance of the cell over the total period of the experiment. In short, a "good" experiment would entail a high degree of precision and accuracy in measurement. The better replications would be ones with more accurate and precise systems of measurement (adequate calibration, sensitive detectors, suitable controls, etc.).

The appeal to standards of a "better" cold fusion experiment helped to sanction less charitable interpretations of Fleischmann and Pons's claims. Fleischmann and Pons, for instance, had used open cells that allowed the evolved gases to escape into the atmosphere. As a consequence, critics argued that their measurement of excess heat was imprecise. They could have no idea, for instance, how much energy might be due to the recombination of gases in their cells. A number of the experimenters at the APS meeting presented data obtained using closed cells, and none of these reported measurements of excess heat.

It is important to note, however, that the use of closed cells could not be seen a priori as a "better" way to do the experiment. Some CF researchers pointed out that their measured heat excess far exceeded the energy that might be generated through recombination of gases, so that the risk of using closed cells (where evolved gases would be under pressure) would hardly be worth it if one was simply attempting to reproduce the phenomenon, especially since Fleischmann and Pons had warned of potential explosions. These researchers suggested that the better experiment would concentrate on reproducing a high magnitude of heat excess so as to get around the need for more dangerous and difficult-to-manage closed-cell calorimetry. This kind of counterargument does not establish once and for all that closed cells are necessarily worse or better.

What it suggests is that standards for what counts as a better or worse way of conducting experimental replication rely on a degree of collective agreement and acceptance of those standards on the part of would-be replicators who put the standards into practice.

## THE RHETORIC OF SIMILARITY

A few of the APS presentations stand out as being different from the range of unsuccessful experiments that were represented as negative replications. An understanding of the relative accuracy and precision of measurement in two contradictory experiments might sway scientists' perception of the reality of cold fusion, but a powerful rejoinder to the most precise negative replications is that these experimenters are precisely measuring nothing. According to this argument, a negative replication, no matter how sophisticated the instrumentation, is not a successful replication because there is nothing there to measure. Thus if experimenters are unable to produce the conditions that might generate the cold fusion process itself, it doesn't matter how sensitive the instruments are; no anomaly will be detected.

The obvious solution to this problem for experimenters like Lewis claiming disconfirmation is to copy the original experiment as closely as possible. A replication that arguably uses the same apparatus and protocol as Fleischmann and Pons and produces positive results using imprecise instruments and negative results using precise instruments would be more convincing. I say "arguably" here because it is impossible to copy an experiment exactly. An exact copy would have to be produced under the same laboratory conditions (if not in the same laboratory) and with the same materials as the original. Consequently, the question of whether an experiment constitutes a copy is a matter of its representation and interpretation. A replication may not be an exact copy, but it can be represented as an exact copy. The reproduction of experiments in this sense is a representational game (Shapin 1984), one that Lewis played with great effect, for instance, in the published version of his APS presentation.

Lewis used a variety of rhetorical means to represent the Caltech experiments as being very similar to those of Fleischmann and Pons. They go beyond the mere statement that the two experiments are similar in design. More subtle invocations of similarity are built into the way the paper is written. Obvious means of conveying the resemblance between the two experiments include frequent referencing of the original Fleischmann, Pons, and Hawkins paper. There are approximately ten references to the original experiment in the six pages of the Caltech paper (as opposed to only two references in a positive replication from Texas A&M, for example). Statements of the following kind are typical of the Caltech paper: "We found a temperature differential of up to 2 C in an unstirred operating cell . . . of dimensions very similar to those used in previous work [reference to the Fleischmann, Pons, and Hawkins paper]" (Lewis et al. 1989, 528). The representational accomplishment here is to allow Lewis's experiment to stand for Fleischmann and Pons's experiment. Lewis could not be present during Fleischmann and Pons's experiment, and as a consequence he could not credibly say what, if anything, they might have done wrong. Yet by representing his experiment as "the same" as the original, Lewis is able to bring Fleischmann and Pons's experiment into his own laboratory to see where mistakes may have been made.

One of Lewis's conclusions was that measurements of excess heat are due to the "inadequate mixing of the various sources of heat in the cell" (Lewis et al. 1989). Fleischmann and Pons's cell had no magnetic stirrer. When Lewis's team added a stirrer to its copy of Fleischmann and Pons's cell, the excess heat disappeared. For Lewis the source of excess heat in Fleischmann and Pons's cell was temperature gradients that built up in the cell. Fleischmann and Pons must have placed their thermistors in a "hot" spot. In this way, Lewis's rhetoric of similarity contributed to a decline in charitable interpretations of Fleischmann and Pons's results.

## The Regress Continues

The APS meeting was an important turning point in the controversy. The interactions of prior social networks combined with the articulation of standards of experimental competence and the rhetoric of similarity to produce an atmosphere of skeptical solidarity in which it became easier for scientists with null results to voice uncharitable conclusions, thereby producing negative replications. This was the beginning of the end for cold fusion, but it was still not enough to kill the claims outright. Fleischmann and Pons were not at the APS meeting, but they did fight back in a press conference and at the annual meeting of the Electrochemical Society held a week later (on May 8, 1989): "Responding to assertions by Lewis . . . that their temperature measurements were flawed, Pons and Fleischmann showed a video recording of one of their cells in action, in which obvious and vigorous bubbling was adduced as the means by which a uniform temperature was maintained by the cell. Furthermore, a dye . . . introduced into the electrolyte was visibly well mixed in about 20 seconds" (Lindley 1989). Fleischmann and Pons denied the similarity claims of Lewis. In the original experiment there *was* adequate stirring. The argument was that Lewis had made a "bad" copy, and Fleischmann and Pons tried to turn Lewis's purported negative replication back into a mere null result.

Fleischmann and Pons also spoke to the issue of precision and accuracy, arguing that their calorimetry was good enough and that a more thorough paper would be forthcoming. Yes, the excess heat and nuclear ash were not explainable in terms of conventional fusion processes, but there was simply more energy being produced than could be accounted for by any and all possible chemical and physical processes. It was not necessary to have precise measurements because the production of excess heat was well beyond any conventional energy balance for the cell.

After these arguments, it was time for the counterattack; this was supplied by Robert Huggins. At the Electrochemistry Society meeting, Huggins criticized Lewis's work, arguing that "it's hard for him to see the effects because he's doing it wrong. . . . the key is making sure that the crystalline structure of the Pd is such that the deuterium atoms can be loaded inside."[27] For his part, Lewis responded, "There's no question that our rods were loaded, there's just no fusion going on."[28] Lewis also reemphasized the similarity of his replication to the original: "Everything was right with that experiment. We did the pre-treatment; we had exactly the same amount of palladium, the same electrode geometry, the right electrolyte, and the right current density."[29] But Huggins and other CF supporters continued to press their arguments. Lewis and others who had not measured the CF effects had failed to do so because they were continuing to do their experiments improperly. Lewis had not run his experiments long enough, he did not use the right current, he did not prepare his cathodes properly; he simply didn't produce the correct material conditions for the cold fusion reaction (whatever it was) to take place. The implication was that the negative replications reported at the APS meeting were simply bad experiments.

Did Lewis prepare his cathodes properly? Did Fleischmann and Pons have adequate mixing in their experiment? As one researcher wrote of his own CF experiments in 1996, "I can offer a nearly useless experimental observation: In our attempts to explore the Piantelli experiment, we heated a Ni rod in a H2 atmosphere to just above the Debye temp and 'soaked' it there for days while applying various stimuli to the lattice (mechanical shocks, magnetic pulse, etc.) and we could observe no sign of excess heat generation. I deem this observation 'nearly useless' because, until we know whether or not CF is real, all negative experiments can be successfully argued to be due to procedural errors."[30] Here, this scientist succinctly states the condition of the experimenter's regress in attempting to defend positive CF results from potential negative replications (including his own). Even Lewis publicly expressed a lack of certainty in this strict sense: "We will never know if the inability of most of the laboratories to do this means that they're doing it very right and Pons and Fleischmann are doing it wrong or if the opposite is true."[31] How then was the regress to be resolved so that a constant uncertainty in principle would give way to an ability to decide correct from incorrect experiments in practice?

At the end of May, the U.S. Department of Energy, in cooperation with Los Alamos National Laboratory (LANL), held a workshop on cold fusion. There were over five hundred participants, and the talks were broadcast live via satellite to many more hundreds of scientists. The purpose of the conference was to try to reach a consensus on cold fusion, but the outcome was anything but a consensus, as the meeting was split between papers reporting positive results and those reporting negative results. Norman Hackermann, one of the co-chairs of the meeting, had this summary: "We have reached no consensus. Those who didn't believe in the phenomenon have not changed their minds. Those who do believe there is something there haven't changed their minds."[32] By the end of May and through the summer of 1989, scientists were not changing their minds but in effect "digging in" as supporters or critics of cold fusion, and those who wished to be neither found it increasingly difficult to take the middle ground.

If no one was changing their minds, how did the controversy end? By the end of 1989 there was certainly a long list of negative replications, but there was also a significant list of positive replications. It was clear that while the majority was against them, Fleischmann and Pons did not stand alone, nor was their group of supporters particularly small or lacking in credibility. In the United States alone, significant support came from well-known researchers at Stanford, Texas A&M, MIT, and the Los Alamos and Oak Ridge National Laboratories, amongst others. The situation seemed to be a standoff. As I have argued, the demise of cold fusion was closely linked to a shift in the degree of interpretive charitability would-be replicators were willing to extend to positive experimental claims. The APS meeting and public displays of skeptical solidarity made less charitable interpretations more plausible, but as long as Fleischmann and Pons (and other cold fusion supporters) were able to defend themselves and dissociate attempts to turn negative results into negative replications, the case for cold fusion might be kept alive indefinitely.

As long as our understanding of closure depends on agreement or near consensus among scientists, we cannot say that the cold fusion controversy has ended. But in the absence of agreement, other closure mechanisms come to the fore. As one critic observed, "Until someone goes to Pons' and Fleischmann's lab and measures what is going on there, you can always say that something has been left out."[33] This comment suggests two successful strategies for bringing the controversy to an end. The first is to attempt to get into

Fleischmann and Pons's lab by obtaining direct access to the original experimental data, looking over their shoulders while the experiments are going on, or barring that, improving the representational similarity of replications done outside the lab. The second and more direct approach is, in a sense, to get rid of Fleischmann and Pons's lab altogether. The first strategy aims at displacing the issue of competence from experiments to experimenters, thereby threatening the experimenters' credibility and their ability to defend themselves. The second strategy removes the material basis for the production of positive results through the deprivation of necessary resources like money, equipment, and labor. What both strategies ultimately accomplish is to remove the capacity for cold fusion supporters to defend themselves in the face of skeptical solidarity.

## *Similarity and the Collapse of Credibility*

One of the most important pieces of evidence for fusion in Fleischmann and Pons's experiments was the measurement of gamma radiation. This was accomplished indirectly, through the interaction of neutrons and the water bath in which the electrochemical cell was immersed. Using a scintillation detector, the gamma rays produced in the CF experiment were identified with an energy of 2.2 MeV. This kind of gamma radiation is not found in nature, and more importantly it is consistent with the 2.45 MeV neutrons that would be produced during normal D-D fusion. The gamma ray spectrum presented in Fleischmann, Pons, and Hawkins's paper had a distinct peak around 2.2 MeV and was a crucial piece of visual evidence for CF.

Problems with the peak began at the end of March, even before it had been published. As noted above, Fleischmann traveled to Europe right after the Utah press conference and gave talks at the Harwell Nuclear Research Laboratory near London (March 28) and at CERN (March 31). At Harwell, Fleischmann showed his gamma-ray data with the crucial peak at 2.5 MeV. The Harwell scientists immediately pointed out that the peak was in the wrong place for indicating neutron capture from D-D fusion. At his talk at CERN the peak was in the "right" place at 2.2 MeV, as was the published peak. While this inconsistency may seem relatively minor, it became the cause célèbre of Frank Close, a physicist and popular science writer, who began to be suspicious of Fleischmann and Pons's ethics in the presentation of their data. Was Fleischmann switching the peak around to suit his argument? Perhaps Fleischmann simply didn't understand enough nuclear physics to recognize what he was doing wrong? Close has been relentless in his criticism in a book, in public talks, and in several short articles.[34]

More damaging than Close's criticism, however, was a short paper published in *Nature* by Richard Petrasso and his colleagues at MIT (Petrasso et al. 1989). This was followed by a response by Fleischmann and Pons (Fleischmann, Pons, and Hoffman 1989), and a final reply from Petrasso (1989). Petrasso criticized the published peak as being incomplete and improperly calibrated. Using a broader spectrum taken from a detector in the background of a television broadcast made in Fleischmann and Pons's lab and an analysis of the published data, the MIT group argued that the published peak at 2.2 MeV was both the wrong size and shape to be the "fusion signal" peak. The published peak was too narrow, and it was missing a feature known as a Compton edge. Petrasso's conclusion was that the 2.2 MeV peak must be an instrumental artifact of some kind.

Fleischmann and Pons's response was interpreted in the media as a less than honest retraction of their gamma-ray measurements, and it is not hard to see why. They attempted to counter Petrasso's deconstruction of their peak by arguing that the spectrum taken from the television broadcast was inaccurate; Petrasso did not have a good copy of their original peak. Moreover, Fleischmann and Pons's detector was not sensitive enough to make definitive judgments about the location and intensity of all features of the spectrum, so that the resolution was not good enough to support Petrasso's analysis. Petrasso was inferring information that simply wasn't available. Finally, Fleischmann argued that their peak was produced during an experiment that produced excess heat but the same peak was not present in the normal background radiation of the lab.

Rather than addressing Petrasso's claims directly, Fleischmann and Pons's response highlighted the idiosyncratic nature of their own claims, arguing that something nonartifactual was being measured; they just were not sure what it was. This seemed to give many scientists, especially physicists, the sense that Fleischmann and Pons were hedging by adjusting their interpretation in an ad hoc fashion. In the public context of appearing in *Nature*, the gamma-peak episode proved to be devastating for Fleischmann and Pons's credibility, which had already been suffering in more private contexts since April. A number of scientists I talked to cited the *Nature* debate as a pivotal point in their own thinking about cold fusion that made them resolve to end their CF research pending a more complete disclosure of Fleischmann and Pons's data. For the news media, the gamma-peak episode provided a perfect arena of direct confrontation to report, and, in the absence of new data from other sources, trust in Fleischmann and Pons's ability to evaluate their gamma-radiation data faltered.

The success of Petrasso's argument rested to a large extent on his ability to argue that the peak he was analyzing as an expert in gamma-ray detection was the same as the peak Fleischmann and Pons had measured. Fleischmann and Pons had done markedly better in defending themselves against Lewis on issues of calorimetry, but against Petrasso they were on less solid ground in not being able (or willing) to defend their competence in gamma-radiation detection. Following the episode, Fleischmann and Pons became more circumspect about their nuclear data; placing very little stress on their radiation measurements, they focused instead on the magnitude and reproducibility of their excess heat claims. What is important about this particular episode is the way Petrasso used Fleischmann and Pons's own peak to raise questions about their competence in making radiation measurements. If Fleischmann and Pons had performed their gamma-ray detection improperly, scientists wondered what else they might have done wrong.

## *The Consequences of Extreme Similarity*

One of the most convincing negative replications was performed by Michael Salamon, a physicist at the University of Utah. Salamon received permission from Fleischmann to perform neutron measurements with his own counters in Fleischmann and Pons's lab, where there were supposed to be working CF cells (defined as cells producing excess heat). After some weeks Salamon concluded that no neutrons were being produced in Fleischmann and Pons's cells and therefore fusion was not a likely explanation for the production of excess heat (Salamon et al. 1990). This replication was seen as being an especially convincing disconfirmation of CF because Salamon was actually in Fleischmann and Pons's laboratory working with their cells. The *Nature* editorial introducing Salamon's paper represented this in the following way: "The account appearing this week is of special significance. . . . Salamon and his colleagues . . . have searched for nuclear emissions from cold fusion cells and found nothing. The difference is that Salamon works at the University of Utah, and the cold fusion cells he examined were in the laboratory of Fleischmann and Pons" (Lindley 1990).

Upon publication of Salamon's results in *Nature*, Pons reacted in a way that many thought inappropriate. He directed his lawyer, Gary Triggs, to send a memo to Salamon demanding that he print a retraction or else face being sued for libel. A letter sent by Triggs to Salamon lists the following concerns supposedly discussed between Pons and Salamon before the results were published:

1. That the paper is factually inaccurate.
2. That there was apparently some selection of data presented in your paper.

3. That the experiments conducted did not reveal the presence of a particular in situ gamma ray calibration standard.
4. That the date of publication of your letter was engineered for editorial reasons by Nature.
5. That certain questions exist as to whether or not your experiments were predesigned to provide the negative results reported.
6. That serious inconsistencies exist between your published paper and other data circulated by you from the experiments which have been available to my clients.[35]

Pons took Salamon's replication as a personal attack on his credibility; to him Salamon was no longer doing science, but rather was maliciously maligning his integrity as a scientist.[36]

One scientist I interviewed had a similar (though less extreme) reaction to the work of Nathan Lewis. "Nate Lewis didn't exist before cold fusion, his reputation has been created by his hostile attack on cold fusion and his stature has largely been a reward for this attack dog type approach. . . . He's good at some things, but he isn't good at calorimetry or physical electrochemistry, it's not his skill and I know that he didn't entertain a very prolonged effort . . . and the non-observation of such a phenomenon is simply akin to saying that I can't climb Mount Everest, but that doesn't mean somebody else can't."[37] In this criticism and in the relation between Salamon and Pons, the consequences of the rhetoric of similarity become clear. The more similar a negative replication is perceived to be, the less participants are able to blame the difference in experimental results on a difference in experimental method and design. If the experiments are the same and the results are different, then it must be the experimenters that are at fault. This creates the conditions for attacks on the personal integrity and credibility of scientists. In the extreme, it can lead to accusations of fraud (especially against nonscientists), but more likely scientists accuse each other of doing "bad" or "pathological" science. My point is that such accusations tend to be put into play when similarity judgments are made.

Similarity judgments in the case of cold fusion became a means for critics to plausibly argue that Fleischmann and Pons and other CF supporters were not doing proper science, and therefore their attempts to defend themselves might be discredited. Thus in regard to the DOE-LANL meeting on May 23, one physicist wrote the following to one of the meeting's organizers:

> After the presentations at the APS meeting on May 3 at Baltimore by Nathan Lewis, Steve Koonin, Robert Parker and others it should have become clear to any leftover doubters that the experiment of the

> Utah chemists is wrong. . . . If there were ever any reasons to hold a meeting on cold fusion several weeks ago, it seems to me that these reasons are no longer valid. On the contrary, it is important now to end this charade as soon as possible. . . . It provides a forum for crackpot physics, chemistry and magic metallurgy. Were it not for Pons and Fleischmann, no one would have dreamed up such a meeting. Why hold it now that they are thoroughly discredited (incompetence, delusion, fraud according to Koonin, Parker and others including myself)?[38]

The APS presentations of Lewis, Koonin, and Parker are used to discredit Fleischmann and Pons and deny the scientific legitimacy of any future meetings on the subject of cold fusion. As "crackpots" and not scientists, Fleischmann and Pons can no longer claim any right to defend themselves in the context of a meeting (or elsewhere).

## *Finally Closure*

Most studies of scientific controversies depend on an implicit assumption that controversies end once a certain level of consensus amongst participants is achieved. I have been chiefly occupied with exploring the ways in which experimental replication, the dynamics of credibility, and the formation of collective beliefs are related in terms of processes of public representation. In cases where sustained disagreement seems limited to only a handful of individuals, understanding controversies in these terms does not seem to be a problem. In attributing closure in the case of high-visibility gravitational radiation, for instance, Collins writes, "By 1975 a number of scientists had spent time and effort actively prosecuting the case against Weber. Most others accepted that he was wrong, and only one scientist other than Weber thought the search for gravity waves was worth pursuing" (Collins 1981c, 38). The data in the case of cold fusion, however, suggest that the attribution of closure through the identification of some kind of near-consensual collective agreement is not tenable. Despite the production of skeptical solidarity and the decline in interpretive charitability leading to the delegitimation of Fleischmann and Pons, support for cold fusion continued through 1989 and into 1990.

It is certainly the case that by the end of 1989 more participants believed that cold fusion was an artifact rather than a new form of fusion. The Institute for Scientific Information (ISI) published an article in March 1990 with the title "Scientists Vote on Cold Fusion: Their Verdict? No, Not Likely." Out of 141 papers referenced in the Science Citation Index, 27% were "positive," 21% were "neutral," and 52% were "negative." Further, 75% of the positive citations were in the context of theoretical articles; only nine of the thirty-eight "positive" citations came from papers reporting experimental results. "To put it another way," the article stated, "the experimental evidence went more than 4 to 1 against cold fusion" (ISI 1990b).

The problem with this is that other participants in the controversy might not share the ISI's interpretation of what counts as a "positive" or "negative" paper. In the context of a listserv discussion in 1991, one scientist explains the situation in the following way,

> You have to realize that you'd get a different assessment from every person. There are those who claim that there are over 200 positive papers, i.e. 200 experiments say yes, yes, yes to cold fusion. I have been through the list, and out of something like 250 experimental papers, I personally regard about 30–40 as "quality" work. Out of this group, there are about 10 positive ones. Again, just what is "positive" is a matter of taste. The TB [true believer] will race for his/her word processor at the tiniest neutron emission, or doubtful mass spectrum pimple. These 10 or so are mostly excess heat or correlations. This is not much out of 600+ papers, but I can't give a "conventional" (i.e. non-hitherto unknown nuclear reaction) explanation for results of, say, Belzner et al. or even FPALH–90. I personally say that a lot of shoddy work has been done, and ought not to be used to support cnf [cold nuclear fusion].[39]

Even if we accept the ISI's interpretation and accept that the increasing majority of negative papers is indicative of the acceptability of the belief that cold fusion is not real, science-studies analysts are faced with a problem. How many scientists must continue to disagree before we can say that a controversy is over? We can continue to track the belief of scientists as they hole up in their respective networks and attempt to win converts or delegitimate opponents, but in the case of cold fusion at least this seems to be a no-win situation. Does this mean that the controversy is not over?

Perhaps, but during a controversy participants on all sides actively engage in deconstructing, transforming, and enrolling each other's arguments and opinions. Closure can be characterized by the development of an asymmetry with respect to these actions; it results not from agreement but from a change in the distribution of power, where power is the ability to act on one's beliefs in the processes of deconstruction and enrollment. Central to this understanding of the distribution of power is the relative legitimacy of actors and the condition of controversies as complex representational and rhetorical games, but equally important are actors' relative access to material resources for doing science (and resources for producing representations in the first place).

The delegitimation of Fleischmann and Pons and cold fusion was ultimately successful because supporters began to lose access to the means for defending themselves. As Latour says, "Arguing is costly. The equal world of citizens having opinions about things becomes an unequal world in which dissent or consent is not possible without a huge accumulation of resources" (Latour 1987, 70).

## *Resource Deprivation*

Scientists never did reach a consensus about cold fusion, but by the end of 1989 the research effort was certainly slowing down. The vast majority of scientists who had tried to reproduce the results, while not necessarily disbelieving the phenomenon, became increasingly skeptical about whether cold fusion was worth pursuing for pragmatic reasons. On June 15, for example, researchers at the Harwell Nuclear Research Laboratory held a press conference announcing they would stop their work on cold fusion after an exhaustive effort that had produced no positive results. The Harwell researchers gave up not so much because they disbelieved cold fusion or doubted the credibility of Fleischmann and Pons, but because they could no longer justify the expenditure of time and resources without anything positive to go on. Cold fusion started to die as the material and human resources necessary for its continued survival were withdrawn. What cold fusion lost was its life support; in laboratory after laboratory the cold fusion research effort died, and this created something of a feedback loop; the more scientists who stopped their experiments, the more justification there was for other scientists to do the same.

### THE DEPRIVATION OF FUNDING

The single most devastating blow to the cold fusion research effort came in the form of a final report from the DOE Energy Research Advisory Board (ERAB) panel appointed to evaluate the cold fusion claims and recommend policy for federal funding. The panel was cochaired by John Huizenga, who would become a staunch cold fusion skeptic, and its members were twenty-two scientists of varying backgrounds. The report's main conclusions were that "based on the examination of published reports, reprints, numerous communications to the panel and several site visits, . . . the experimental results of excess heat from calorimetric cells reported to date do not present convincing evidence that useful sources of energy will result from the phenomena attributed to cold fusion. . . . The panel concludes that the experiments reported to date do not present convincing evidence to associate the reported anomalous heat with a nuclear process . . . that the present evidence for the discovery of a new nuclear process termed cold fusion is not persuasive."[40] The report's explicit policy recommendations did not seem as severe, however, stating simply that "the panel recommends against any special funding for the investigation of phenomena attributed to cold fusion. Hence we recommend against the establishment of special programs or research centers to develop cold fusion."[41]

The report was not recommending against all funding for CF research, but a closer analysis of the panel's recommendations shows a preference for some kinds of research over others, supporting the position of critics against the arguments of Fleischmann and Pons. Thus one recommendation reads, "A shortcoming of most experiments reporting excess heat is that they are not accompanied in the same cell by simultaneous monitoring for the production of fusion products. If the excess heat is to be attributed to fusion, such a claim should be supported by measurements of fusion products at commensurate levels." (Energy Research Advisory Board 1989, 3). Here the report reinforces conventional practices in nuclear physics by specifying as a condition of funding that claims of fusion be "supported" by evidence of nuclear ash at levels commensurate with excess heat. In this way the report was legislating standards of experimental competence that were in dispute during the APS meeting; the idea that excess heat should be commensurate with measurements of nuclear ash was the very assumption that Fleischmann and Pons were trying to challenge. For Fleischmann and Pons (and increasingly for others), cold fusion was defined in terms of the absence of commensurate heat and nuclear ash, so why should the measurement of these things be a condition for receiving DOE funding?

Some researchers have charged that the panel was biased against cold fusion from the beginning. As one electrochemist told me, when the panel visited his lab, "I remember being very surprised at Huizenga's attitude at that stage in that he was asking questions [as if] he had already decided what he was going to write in his report and he was asking questions to provide confirmation for suspicion. He was completely uninterested in my answers if I didn't answer in the way he wanted me to answer. He was anything except a dispassionate scientist."[42] He continued by saying that

> a number of the cold fusion community question the scientific merit and the findings of the report, but I think the things that were written were largely right and the conclusion—which has been often misquoted—but the conclusion was that insufficient evidence for a real phenomenon exists to merit special funding in this area—that funding proposals should be submitted to DOE and handled in the normal way. . . . that is exactly right. It's the interpretation that's the

> problem . . . because if you are in a position of reviewing a proposal you can decide its fate by who you send it to to review. So if the proposal was sent to Huizenga to review the answer would always be don't fund it and that's exactly what DOE did for three years. They sent all the proposals to Huizenga—gave them the Huizenga test. And if Huizenga said fund it they funded it, and of course he never did.

The decline in interpretive charity and the delegitimation of Fleischmann and Pons were supported by a simultaneous withdrawal of funding for CF research. Here the social dynamics leading to the formation of collective belief represented in the opinion of Huizenga made possible, and was made possible by, the deprivation of funding for defending the counterclaim. As a consequence, scientists wishing to conduct further research on cold fusion ran into substantial difficulty.

When I asked about how he supported his research after 1989, one engineer told me, "I did think of doing the gas-experiments. I would have done it if I had the support but we couldn't get any support. They wouldn't even peer review them [laughs] you'd send a proposal to study this area and. . . . I tried DOE, I tried NSF. . . . They pretty well put a stop to any work—I even called the people there and got a really cold fish answer."[43] Another physicist said of the DOE, "They have not expressed any interest in this area in particular, on the other hand they claim they have no bias against the area. There has been two or three proposals sent to them historically and none of them passed review. The reviews have been pretty harsh, so they expect if they get a review on this area they'll just get chewed by the referees and there'll be no funding. . . . In practice there is no place to go."[44]

Some major funding sources did remain for CF researchers in the United States outside the DOE. The Office of Naval Research (ONR) and other military agencies continued to support CF research at modest levels, and the California-based Electric Power Research Institute (EPRI) funded several researchers reporting positive results up until 1996. Yet despite this, the negative report from the Department of Energy effectively marginalized cold fusion as a concern in either mainstream or alternative energy research, making funding from other private sources more difficult to obtain.

Even in Utah, where support for cold fusion was strongest, things did not go well. In April 1989, the Utah legislature appropriated $5 million to establish the National Cold Fusion Institute (NCFI). Opening on August 7, 1989, the NCFI supported several cold fusion research projects, including that of Fleischmann and Pons. On June 30, 1991, the NCFI was closed down. The original $5 million in seed money had been spent, and the expected partnerships with private industry had not developed. The NCFI also suffered from a scandal in 1990 over a misallocation of funds at the University of Utah, which resulted in the resignation of the university's president, Chase Peterson. The lack of corporate interest in cold fusion could be related in part to a decline in interpretive charity and processes of delegitimation, where the weakening of the prospects for cold fusion meant that investment was becoming an increasingly risky venture, whether one believed the phenomenon was real or not. As one engineer explained the situation to me, "The field still hadn't produced anything big, and looked like it had a lot of problems. You might say it's a big potential payoff field but it also had a big unpredictable factor and you could very easily spend money and never see any return on it. I think that's still true."[45]

### THE DEPRIVATION OF EQUIPMENT, LABOR, AND TIME

An important consequence of the withdrawal of funding opportunities was that it became increasingly difficult for researchers to construct more sophisticated experiments that might help resolve the problems associated with cold fusion. Difficulties included gaining access to instrumentation and even laboratory space. One engineer had this story of his research efforts in 1993: "I originally set up at [a university]. I'm very good friends with a professor in the biology department there. . . . I set up in his lab, he had a tiny lab with some space available and of course that also gives me a bit more in the way of tools and materials. But then last summer, he was retiring and had to give up his lab and I went off for four months. . . . So at that point I disassembled the lab and moved the pieces to my garage, I reassembled it there after I got back. I've done only a few experiments since I got back."[46] In the absence of formal funding, many scientists were forced to draw on the limited resources available from their discretionary funds and personal networks. At the same time, they had to content themselves with conducting research on a smaller scale, and in many cases were forced to stop research altogether.

Scientists working in universities or industrial laboratories found it difficult to get approval for continued CF research from their superiors, and they were often put into a position of doing research on a part-time basis. Another related issue in universities was finding students who were willing and able to do experimental work. As one researcher told me, "There's no students or readily available people. See, you need hands to put anything together, it takes a lot of work to even put together pipes much less instrumentation, check it out, get it working, verify that it's not fraudulent, instruments, etc."[47] In another situation, described at the beginning of chapter 1, the problem of recruiting students was so dire that the scientist was happy to draw on my modest skills in helping to prepare electrolytic cells, cleaning components, and taking measurements. This experience forms the basis of part of my discussion in chapter 6. With these examples we can begin to see the feedback processes that help produce closure in this case: the delegitimation of cold fusion facilitates the deprivation of labor, which in turn facilitates further delegitimation due to the absence of research.

## THE DEPRIVATION OF COMMUNICATION

Latour and Woolgar (1979) have observed that the primary product of scientific work is information in the form of publication. In university science this usually takes the form of scientific papers, and in industrial science it takes the form of patents and reports (Packer and Webster 1996). In addition to the deprivation of the means for producing scientific papers (funding, equipment, and labor), the closure of the cold fusion controversy has also been brought about by the deprivation of the means for communicating scientific work. Again, this may be linked to processes of delegitimation: "Right now my scientific reputation, which once was rather good, is essentially nonexistent. I can't get papers published mostly because my name is on them. I can't get proposals funded. . . . when all is said, and all is done, unless something turns up rather good, I probably won't see another promotion as long as I stay here. I probably won't be successful in getting much of any students, or funding or anything. . . . it's a very, very bleak future right now."[48]

CF researchers believe that one reason their work isn't accepted for publication is their delegitimation, and evidence in support of their perception can be found in reviewers' comments on their submissions. In one case, a reviewer wrote the following about a paper that was eventually rejected for publication:

> It is the type of manuscript we saw in March 1989, but definitely not the definitive piece of work one can expect in 1994. The authors start from a biased point of view and only reference previous works, almost all discredited, which make claims for tritium production in the electrolysis of D2O with Pd cathodes. They do not cite any of the work that shows no tritium production. . . . WHERE ARE THE NEUTRONS? . . . As a former member of the Editorial Advisory Board of [this chemistry journal] I would be offended to see a manuscript of such dubious worth published in [this chemistry journal].[49]

In his comments, the reviewer identifies the paper in terms of cold fusion and its discrediting, referring to a version of cold fusion that requires the presence of neutrons as a definitive test of competence. The author, who was not looking for neutrons, received this review and in his reply tried to dissociate his work from cold fusion so that it might stand on its own merit:

> Please, accept the fact that we examined the behavior of the Pd/D and not the Pd/H system; thus, no need for light water experiments. If possible, dissociate yourself from the notion that we have attempted to prove "cold fusion" (i.e. there is no need to know WHERE ARE THE NEUTRONS?) and consider the data only in the context presented, i.e. you should have read what was written and not what you thought to be our intent. It is quite obvious that you started from a biased position and ended in the same position.

Exchanges like this one have not been uncommon, assuming, that is, that authors are able to get their papers considered for review in the first place.

As with funding and material resources, however, avenues of communication for CF researchers have not been completely cut off. I will address the issue of communication again in chapter 6, but it is important to note that some articles do make it through peer review. Mainstream journals such as the *Journal of Electroanalytical Chemistry*, *Physics Letters A*, *Il Nuovo Cimento*, and *Fusion Technology* have published articles on cold fusion over the past ten years. Thus CF researchers do not so much have the experience of being blacklisted as they have an increasingly difficult and frustrating time organizing and disseminating their research. Closure then should not be understood in this case as any kind of sudden or catastrophic event but rather as an ongoing process involving the choking off of resources essential for the survival of dissent.

## On the Concept of Closure

There is relatively little about the story of cold fusion as I have told it so far that should surprise science-studies scholars. It reaffirms the important epistemological claim that controversies concerning novel experimental phenomena do not reach closure in the logical or rationalist sense of the term. No argument, criticism, or experimental demonstration could serve alone to invalidate the claim that cold fusion is a real phenomenon, since every argument, criticism, and demonstration could be found to have flaws or errors of some kind according to at least some of the participants (Collins 1985, 1989). This does not mean that arguments, criticisms, and experimental demonstrations are not important. As a matter of fact they are crucial, as they remain an essential currency of exchange in the debates and maneuvers that lead to a resolution. The problem is that arguments, criticisms, and experiments alone are not enough to produce closure and end the experimenter's regress.

It is worth reiterating that the epistemological conclusion to be drawn from this observation has never been one of relativism. Although controversies could in principle go on forever, they do not do so in practice. Methodological relativists and constructivists find agreement with their rationalist and realist interlocutors in arguing that controversies do reach closure; it is only how closure is achieved that sparks heated debates between philosophers and sociologists. The story of cold fusion certainly supports a social-constructivist view of closure against a rationalist one, but it also unsettles a social-constructivist view of what closure amounts to.

Closure is a product of participants' social and material maneuvers to establish a collective sense of what will count as a fact and what will not. The cold fusion controversy ended, not because a majority of scientists believed that cold fusion was an experimental artifact, but because it became incredibly difficult for those who believed otherwise to legitimately act on their beliefs. Legitimately acting on ones' beliefs in science means having access to the resources necessary to enact ones' beliefs in a scientific context. Typical resources include funding, laboratories and equipment, skilled labor, time for conducting research, and means of communication (conferences, journals, reports, etc.). In this case, knowledge about the nonexistence of cold fusion was stabilized through the delegitimation of dissenting actors and the deprivation of the resources needed to sustain their dissent.

The extent to which closure by delegitimation and resource deprivation can be generalized remains a topic for further comparative investigation into other cases. An implicit part of the argument of this chapter is that science-studies analysts may mistakenly take the relative silence of dissenters in some controversies as an indicator of consensus and closure. In the case of cold fusion, it is clear that dissenters (the cold fusion supporters) did fall silent, but only in certain contexts; in others they remained just as vocal as ever. The situation may be similar in other controversies as well.

In the second chapter I examined how representations of cold fusion configured the core set of experts (as well as the material conditions) in the controversy over the replication of the CF effects. In this chapter, I looked at the kinds of maneuvers actors deployed in an effort to bring the controversy to an end. In this case, the end was marked by the increasing inability of one group (the supporters of cold fusion) to legitimately act on their beliefs in normal scientific contexts. The next three chapters consider the state of affairs after this point, picking up the events of the story after 1990, in what I characterize as the post-closure period.

# *Chapter 4* The Pallor of Death

## *"The Saint"*

It is sometime in the near future. A massive crowd has gathered in Red Square to hear an announcement from the Russian president. It seems that scientists have finally discovered a way to produce endless inexpensive energy as a solution to Russia's worsening energy crisis and economic problems. The crowd is skeptical, angry, and irritable. It is midwinter, Moscow has been without heat for weeks, and the president has made false promises before. It is clear that there is a lot riding on this moment, and if things do not go well now there could be another revolution.

The president appears and makes a speech about the dawn of a new energy age; he pulls back a sheet, and everyone sees a large mechanical device sitting at the center of the platform. The device has a large chamber with liquid in it. Someone throws a switch, and everyone hears a noise—there are a few moments of dramatic tension—will anything happen? Then, suddenly, the chamber on the device begins to glow. The glow intensifies until a bright blue beam of light rockets skyward. The crowd, in awe, backs away from the platform, feeling the heat coming from the device, and the sound of cheering fills Red Square. The "infinite energy" age has begun.

The energy-producing device in the middle of Red Square is a large cold fusion cell, but this scene is not "real"; it marks the climax of the 1997 action-adventure film *The Saint*. Based on the 1960s television series of the same name, the movie is the story of how a spy named Simon Templar (played by Val Kilmer) helps an American scientist named Emma Russell (played by Elizabeth Shue) establish the reality of cold fusion against a background of general scientific skepticism and a plot by a Russian crime lord to steal the secret. The moment of truth (scientific, political, and moral) occurs in the scene I have just related. All doubts disappear with the very public demonstration of a cold fusion device emitting a beam of heavenly blue light that can heat Red Square. At the end of the film, Emma releases her discovery freely to the world and is honored by her fellow scientists. Through Emma's dedication, cold fusion (and Fleischmann and Pons, who are credited with its discovery in the film) emerges triumphant. In Emma's words, "Sure, cold fusion's had a difficult childhood, we're orphans, bastards at best. . . . But difficult childhoods make interesting adults, don't you find?"

The film's reference to cold fusion is surprisingly direct and explicit. Indeed, what *The Saint* lacks in terms of an original plot line (and good acting) is more than made up for by its representation of science and its creative blurring of history and fiction. Although it is set in an implausibly xenophobic near future, there are important elements of the film that refer directly to the cold fusion controversy of 1989.[1] Early in the film, for instance, Emma is seen giving a talk to a group of students at Cambridge University, where she explains the basic physics of fusion (and cold fusion) holding a very accurate replica of Fleischmann and Pons's electrolytic cell. During the course of the film, Emma emerges as the one scientist who has been able to solve the problems that other cold fusion researchers in 1989 could not.

While the film is understandably silent about how Emma Russell technically manages this feat,[2] it is remarkable in its attention to the details of what could otherwise be construed as an obscure piece of scientific history. The attention certainly testifies to the popularity of the controversy in 1989, but it also adds new layers of meaning to the understanding of cold fusion for both the lay public and the film's potential audience of scientists.[3] As a medium of public representation (like a newspaper report or a journal article), the film constructs cold fusion as real, but this construction occurs in the context of a fictional future as opposed to the actual present (or past). The appearance of cold fusion in *The Saint* is evidence that it has not disappeared, and yet, set against the background of the controversy that precedes the film, what is being configured is not cold fusion as a scientific fact but rather cold fusion as science fiction. What sort of strange transition has taken place?

Like the media reports and the journal articles before it, *The Saint* configures cold fusion as a particular kind of object and cold fusion researchers as particular kinds of subjects. The film defines an identity for cold fusion and cold fusion researchers alike. The claims of cold fusion were spectacular to begin with, but they were credible because reputable experts were making them. Mediated through expert voices in March and April 1989, a phenomenon that seemed incredible became credible and enabled a scientific investigation on an unprecedented scale. In 1997, a reversal had taken place as what was once credible became literally incredible to the point of being pure fiction, Hollywood style. This is where cold fusion appears to be most visible to the majority of us today; not in scientific journals or as heating devices in our homes but, at its best, as an element of science fiction and, at its worst, as pure fantasy.

## *Black-Boxing Cold Fusion*

The majority of social and historical studies of scientific controversy have focused on accounts of success rather than failure. Although large-scale public controversies are rare in the history of science, they are usually seen as moments of significant change or revolution (Kuhn 1970). The aftermath of these controversies is figured in terms of accounts of the normalization, routinization, or institutionalization of the "agreed upon" facts. As noted in chapter 1, this process is sometimes referred to as "black-boxing" in the science-studies literature. This is the process by which contested facts become "well-established" and "unproblematic" as they are taken up in routine scientific practice, as noncontroversial results in journal articles, as established facts in textbooks, as unproblematic elements in standard procedures, laws, or formulae, and as components of further material realization in instruments or devices. An important reminder here is that scientific claims do not become a part of textbooks or routine practice because they are black boxes (i.e., because they are unproblematically true). Rather, the seamless integration or knitting together of scientific claims, practices, textbooks, and institutions constitutes the process of black-boxing. This is the scientific process that allows us to understand claims as being unproblematically true (Latour 1987).

For Latour, black-boxing becomes a measure of the "reality" of a claim. Cold fusion would be real, for instance, if "it" could circulate through networks of scientific practice (and beyond) without being challenged, or without encountering significant resistance. This is what became of Müller and Bednorz's claim of high-temperature superconductivity. The spate of initial replications was successful, and the existence of $HT_c$ quickly became a normal feature of routine scientific, and later public, life. What was once the unconventional claim of two scientists has now become the key background assumption in a lucrative field of research and has been embodied in ceramic materials that can make high speed trains go faster and that increase energy efficiency in electronics. Tracing the route from unconventional claim to faster trains, accounting for black-boxing in science, has become the sociologist's methodological equivalent of epistemological justification. But instead of looking for rational grounds for believing that this or that statement is true, we look at the strength and resiliency of the social and material networks through which a statement circulates. We can now understand and even measure scientific change by following the progress of facts as they move from their controversial or problematic beginnings to their position as taken-for-granted facets of modern technoscientific networks.

Consequently, an unjustified or false claim is a claim that is not able to circulate through networks, and does not get black-boxed. In this sense we might argue that cold fusion didn't die but rather never got the opportunity to live (Latour 1996). From the beginning, cold fusion required the careful attention of experimenters (amongst others) in order to live, in order for the claim to be sustained. We might say that it was stuck on life support, unable to exist on its own in other scientists' laboratories or outside them. Cold fusion remained tethered so very tenuously to the apparatus in the University of Utah laboratory. It needed the steady support of Fleischmann and Pons and their allies, as well as the attention of journals, funding agencies and the patent office. For a few months it was touch and go, but in the end the plug was pulled. Such an interpretation lends credence to the idea that the announcement of cold fusion was premature; perhaps it simply was not ready to survive.

Latour views success or progress in science in terms of a hegemonic extension of networks that allow objects to circulate without resistance. This idea also gives us a definition of failure in science: a severing of the network, which results in an inability of objects to circulate. "Every time you hear about a successful application of a science, look for the progressive extension of a network. Every time you hear about a failure of science, look for what part of the network has been punctured" (Latour 1987, 249). Seen from this point of view, the story of cold fusion is a story of scientific failure, because at least as of July 2001 (as the manuscript for this book enters the final phase of revisions), the phenomenon has been unable to circulate unproblematically through technoscientific networks. Fewer and fewer scientists are working on the topic, experimental replication has been inconsistent even for those who claim success, it has been difficult to publish research in peer-reviewed journals, applications for research funding and patent claims have been rejected, and demonstration devices have failed to reach scientific audiences. It is the accumulation of all this that makes cold fusion into something less than real and cold fusion researchers into scientists who are less than credible.

From the arguments of the previous two chapters it may be possible to develop cold fusion as a kind of failure story in the manner described above, but I have something else in mind. Drawing on the same principles that model scientific success as a consequence of the unproblematic circulation of objects through networks, what we can begin to notice is that while one version of cold fusion continues to have trouble getting off the ground, there is another version that circulates rather widely and without much difficulty. This is the cold fusion of *The Saint*, and I will argue that it is at odds with the cold fusion still being investigated in laboratories around the world today.[4]

Indeed, what I will argue in this chapter is that in the case of cold fusion the closure of the controversy did not result in the end of the circulation of cold fusion, but rather in its reconfiguration, transformation, and recirculation as an object of fantasy and pathology. The outcome of the controversy is not simply the belief that cold fusion does not exist, but more to the point, that cold fusion is *not real*. That is, it positively belongs to a category of objects that are not real. Latour's actor-network perspective makes this curious slippage possible. What ceases to circulate in 1990 is cold fusion as a real scientific object, but it does not disappear. Instead, it becomes an example of an unreal scientific object, or more precisely, an object of science fiction or worse, as a pathology of science. The conclusion of the controversy thus results in greater social-epistemological consequences than simply the disproof of cold fusion. In being disproved, cold fusion becomes really not real.

Taking its cue from *The Saint*, this chapter presents the argument that the closure of the controversy accomplished by 1990 did not have as its consequence the elimination or repression of cold fusion as an object of science. In the wake of cold fusion's death, both the scientific community and its publics circulated accounts of the controversy's history. These accounts are tied to practices that continue to produce cold fusion not as real but as really not real. If part of understanding the character of closure (and hence of scientific knowledge) is noting actors' deployment of claims as black-boxed or unproblematic statements of fact, then the character of closure in this case involves the unproblematic circulation of the statement "cold fusion is not real" as fact. The black box, in this case, is the coffin of cold fusion.

But instead of being buried, the coffin of cold fusion remains visible and on parade as an object to be circulated and deployed by scientists amongst themselves and by a heterogeneous group of people that might be called, following Gieryn (1983), epistemological boundary workers. This group can be identified by the public discursive labor they perform in marking the limits or boundaries of legitimate science (what kinds of knowledge count as science, and who counts as a scientist) in different contexts. Examples of epistemological boundary workers include philosophers of science like Karl Popper, scientists acting as popular-science writers and journalists like Robert Park and Martin Gardner, scientific organizations like the American Physical Society, broader "grass-roots" organizations like the Skeptics Society, and government agencies that establish regulatory and funding guidelines explicitly including some kinds of knowledge as science and excluding others.

As I shall argue below, the boundary work performed on cold fusion accomplishes four related things in terms of the production of scientific knowledge. First, as has already been discussed in the literature on boundary work, it protects the cognitive authority of science by differentiating what is to be recognized as good from bad science. Second, while mainly performed outside the locations where most scientific work occurs (laboratories, journal articles, grant proposals, etc.), boundary work constitutes an object that is nevertheless deployed by scientists in routine scientific practice. Third, the production and circulation of this object reinforces closure (and stabilizes knowledge) through the generation and maintenance of a kind of epistemological garbage pail for dealing with future anomalies in energy-related research. And fourth, closure is further reinforced by the continual undermining of the legitimacy of cold fusion researchers in their attempts to reopen the controversy.

## *From the Demarcation Problem to Boundary Work*

The discussion of how science might be different from other ways of knowing about the world has a long and turbulent history, especially in the philosophy of science. The project of understanding what makes scientific knowledge different and why science provides a better approximation of the truth than other kinds of knowledge making can be seen as part of the very raison d'être of the philosophical subdiscipline epistemology. Amongst philosophers the issues are embedded in what is known as the demarcation problem: what, if any, are the epistemological criteria for recognizing and adjudicating the difference between scientific, pseudoscientific, and nonscientific knowledge claims? In the twentieth century alone, the question has been a central concern of philosophers of science like Carnap and Reichenbach, Popper, Lakatos, and Feyerabend, and a significant number of others whose attempts to resolve the issue fill hundreds of pages in philosophy journals.

While debates over demarcation are quite old, much of philosophers' concern over the epistemological status of "scientific" knowledge in the contemporary sense of the term can be traced to the early-modern institutionalization of natural-philosophical practices (for example in the Royal Society in seventeenth-century England). In this period, there was little disciplinary division of labor between those one might call epistemologists and those one might call natural scientists. As a result, solutions to the problem of demarcation were part of the social and political legitimation of the cognitive authority of the natural philosophers as a group.

Studies of early-modern natural philosophy, such as Shapin and Schaffer's (1985) work on the controversy between Robert Boyle and Thomas Hobbes over the interpretation of early air-pump experiments, have shown the seamless relationship between laboratory practice and its legitimation in the form of what Taylor has called the "rhetoric of demarcation" (Taylor 1996, 130–134). Indeed, Shapin and Schaffer's argument hinges in part on being able to show how the development of Boyle's experimental practice and his ability to produce "matters of fact" was dependent on articulating a special difference between Boyle's kind of knowledge claims and the knowledge claims of others (as well as the means for evaluating those differences). For the natural philosophers, demarcation was as much a practical social and political problem as an epistemological one. Success in demarcation meant nothing less than the establishment of the disciplinary foundations for the development of modern science, its practices, institutions, and forms of representation.

Since the development of the division of labor between philosophers of science and professional scientists, the problem of demarcation has gone in different directions. Philosophers have almost refined the issue of demarcation to a science in its own right, developing various frameworks for establishing the "scientificity" of truth claims irrespective of how scientists themselves might treat those claims. Post-Kuhnian epistemology has been rather less taken with the problem of demarcation, however. This is mostly because criteria of demarcation interfere with philosophers' concerns about the epistemological warranting of knowledge claims, whether they are scientific or not (Laudan 1983).

From scientists themselves, however, issues of demarcation are receiving renewed attention, although not in the way that saw Robert Boyle mixing elements of experimental virtuosity and the politics of demarcation in the same representational contexts. Issues of demarcation arise for contemporary scientists less in the context of their normal practice and more at the interface between science and the rest of society. The scientificity of knowledge claims gets worked out not in the laboratories (since work that occurs there is already presumed to be scientific) but in the more public contexts of keynote addresses, newsletters of professional associations, newspaper and magazine articles, popular books, congressional hearings, and courtrooms.

Ostensibly the goal of demarcation work is no longer the legitimation of the epistemological authority of science, but rather the protection of its legitimacy through a policing of disciplinary boundaries in an attempt to determine what kinds of claims and practices count as scientific. Thus issues of demarcation arise, for example, in conflicts between evolutionists and creationists, between the American Medical Association and practitioners of alternative medicine, and between scientists and astrologers. If these conflicts appear in scientific journals, they are confined to editorials and letters, but seldom if ever do boundary conflicts arise explicitly in technical papers.

The articulation of demarcation by scientists has been identified by sociologists of science as an important part of the scientific enterprise. As Gieryn has argued, "Demarcation is routinely accomplished in practical, everyday settings: educational administrators set up curricula that include chemistry but exclude alchemy; the National Science Foundation adopts standards to assure that some physicists but no psychics get funded. . . . Demarcation is not just an analytical problem: because of considerable material opportunities and professional advantages available only to 'scientists,' it is no mere academic matter to decide who is doing science and who is not" (Gieryn 1983, 781). Gieryn called boundary work "the ideological efforts by scientists to distinguish their work and its products from non-scientific intellectual activities" (782). In his initial treatment of the issue, Gieryn associated boundary work with a kind of public rhetoric in which "scientists describe science for the public and its political authorities, sometimes hoping to enlarge the material and symbolic resources of scientists or to defend professional autonomy" (782).

In the first instance, Gieryn interpreted boundary work in somewhat functionalist terms, a means for science as an institution to protect and further its cognitive authority. He later refined this interpretation so that boundary work becomes a discursive strategy actors employ to legitimate and delegitimate scientific knowledge claims in confrontational situations: "Boundary-work occurs as people contend for, legitimate, or challenge the cognitive authority of science—and the credibility, prestige, power and material resources that attend such a privileged position. Pragmatic demarcations of science from non-science are driven by a social interest in claiming, expanding, protecting, monopolizing, usurping, denying, or restricting the cognitive authority of science" (Gieryn 1995, 405). The view of boundary work in terms of legitimating ideology or public rhetoric has provided one way, for instance, for science-studies scholars to redeploy Merton's work on the ethos of science (Merton 1973). Mulkay first suggested that Merton's norms of communalism, universalism, disinterestedness, and organized skepticism might not be conceived as functional constraints on scientists' behavior but rather as rhetorical "vocabularies" or resources for ideological descriptions of science (Mulkay 1979, 71–72). Taylor has more recently taken up Merton's norms as rhetorical resources for boundary work in brief analyses of the creationism debate and the cold fusion controversy (Taylor 1996, 177).[5]

More recently, Gieryn (1999) has taken to looking at boundary work as a practice of cultural cartography aimed at demarcating, expanding, and protecting regions of cultural space (within and outside science) in terms of their epistemic authority. In this sense scientific boundary work marks epistemic territory that defines the legitimate methods, practices, and speakers within that space that constitute "real" science as opposed to practices that pretend to science. These are categories of posers enumerated by Gieryn: pseudoscience, amateur science, deviant or fraudulent science, bad science, junk science, and popular science (Gieryn 1999, 16). We should now also add pathological science, cargo-cult science, and voodoo science to the list. In my view the move to more explicit cartographic metaphors for understanding boundary work highlights the ways in which what counts as science is a product of these public and continuous demarcation exercises. Science gains epistemic authority through its ability to demarcate itself from epistemic practices that are not proper science. Thus cold fusion has come to play an important cartographic role as the Gulag of contemporary technoscience.

## *Boundary Workers and the Representation of Cold Fusion*

The prevalence of public boundary work after 1990 marks a tangible shift in the representation of cold fusion. While references to cold fusion were disappearing in the specialized forums of mainstream journals and conferences, they were appearing in more general public contexts, in popular books, science magazines, public lectures, newspaper articles, and university course syllabi. What was occurring was a form of boundary work in Gieryn's original sense of the term. The story of cold fusion was being retold as a lesson in how not to do science, ostensibly as means of reinforcing or reinstantiating the cognitive authority of science. Consider the following statement:

> Cold fusion is an example of bad science where the normal rules and procedures of the scientific process were violated. One can only be amazed by the number of scientists who reported confirmation of cold fusion by press conference, only to follow later with a retraction or at least a confession of irreproducibility. It has taken upwards of some fifty to one hundred million dollars of research time and resources to show that there is no convincing evidence for room temperature fusion. Much of this effort would not have been necessary had normal scientific procedures been followed. The idea of producing energy from room temperature fusion is destined to join N-rays and polywater as another example of a scientific aberration. (Huizenga 1992, 235)

This comment is from Huizenga's *Cold Fusion: The Scientific Fiasco of the Century*, a book published in 1992. Huizenga, then a professor of chemistry and physics at the University of Rochester, was invited by Glenn Seaborg, the U.S. secretary of energy, to be the cochair of the DOE ERAB panel appointed to evaluate claims of cold fusion during the summer of 1989, as noted above. Huizenga's book is an account of the controversy intended primarily for a popular audience, and his is just one of several such books published in North America and Europe. Others mentioned before include *Too Hot to Handle: The Race for Cold Fusion* (1991a) by Frank Close, a theoretical physicist and science writer, and *Bad Science: The Short Life and Weird Times of Cold Fusion* (1993) by Gary Taubes, a science writer who regularly contributes to the AAAS journal *Science*. Similar characterizations of cold fusion can be found in Friedlander (1995), Dewdney (1997), and Park (2000). There are some important exceptions to the characterization of cold fusion as bad science, most notably Mallove (1991), Fox (1992), and Beaudette (2000), but in general these authors have been fighting a losing battle over the public representation of cold fusion.

The texts of Huizenga, Close, and Taubes all offer accounts and analysis of the events that occurred during the controversy in 1989 and 1990. They relate most of the scientific criticisms of Fleischmann and Pons's experiments and the early replications and are rich with first-person narratives from experts giving their opinions as to the credibility of the various claims. All of these texts are extremely critical of the manner in which Fleischmann and Pons presented their data, arguing that there was never any convincing evidence for cold fusion. Moreover, what little evidence there was had been refuted during the summer of 1989. With this point made, what remains for these authors is to account for the occurrence of the controversy in the first place. How could so many scientists become so excited about a phenomenon for which there was no reliable or convincing evidence? How did things get so out of hand? Close blames the unprofessional behavior of Fleischmann and Pons: "The seeds for the disaster were sown when the first errors in measurement were pointed out to Fleischmann and Pons: by attacking when they should have retreated they set themselves on a dangerous course. They should have withdrawn their nuclear 'evidence' and insisted on the heat, which may have some scientific substance. . . . Instead, their strong belief that they had more evidence than in fact they had, led them to present data that had been obtained more by enthusiasm than by careful science. . . . They are victims of their own excessive claims" (Close 1991, 349).

Close's explanation is formulated in hindsight. Now that the controversy is over and the errors have been exposed, we can say that Fleischmann and Pons "should" have acted differently. Close's book (and indeed all of the others as well) is written in what Pickering (1989) has described as a "realist mode" in which the truth of the matter is treated as given and what is required is an explanation of how the truth was uncovered. The logic of this realist stance necessarily gives way to questions of how respected scientists like Fleischmann and Pons could have been so wrong, why they were not willing to admit their error, and how so many people (scientists and nonscientists) proved to be so gullible.

Thus Huizenga introduces his last chapter by asking, "How did cold fusion germinate and what fueled the whole episode while reaching such gigantic proportions? Was the dream of limitless, clean and cheap energy so powerful that it bred error and self-deception?" (Huizenga 1992, 215). Since cold fusion has already been disproved by the due process of science, the belief in cold fusion both during and after the controversy requires an explanation. Huizenga proposes that it is a consequence of "bad" or incompetent scientific research and that others got carried away because the biases of media reporting, institutional politics, and the personal desire for fame and fortune interfered with the normal process of scientific investigation.

At the same time, however, the explanation of cold fusion in terms of incompetence is somewhat problematic for Huizenga. If Fleischmann and Pons were trained and accredited as scientists and they turned out to be incompetent, then perhaps other scientists are guilty of such incompetence as well. Interpreted this way, Fleischmann and Pons's incompetence becomes a threat to the collective integrity of science itself.[6] Huizenga's solution to this problem is to deploy Langmuir's concept of "pathological science" in an effort to argue that sometimes scientists become blinded to reality and cannot admit their mistakes because of external social and psychological pressures. Fleischmann and Pons are not bad scientists so much as they are suffering from a peculiar malaise that keeps them from behaving properly.

Importantly for Huizenga, the fault does not lie with science as an institution or with Fleischmann and Pons's training or credentials, but rather with a few misguided individuals who are not acting as scientists "normally" do, in part because the social pressure to succeed distorts their views. Indeed, portrayed in this way, the failure of cold fusion is turned into a success for the inherent error-correcting mechanisms of science: "The general scientific enterprise is vibrant and healthy and has weathered the cold fusion flurry with only minor bruises and scratches. The cold fusion fiasco illustrates once again, as N-rays and polywater did earlier, that the scientific process works by exposing and correcting its own errors" (Huizenga 1992, 236).

While historians and sociologists of science have been generally critical of the realist and asymmetrical explanatory perspective of Close, Huizenga, and Taubes, at least one noted historian of science, Daniel Kevles, agrees with

Huizenga and redeploys the point about science as a self-correcting enterprise in a review of Taubes's book: "Whatever its source, the fiasco was not exposed by governmental investigations into scientific fraud but by the relentless scrutiny of other scientists. Cold fusion was not only a story of bad science but a parable of good science—of respect for true evidence, and the courage to admit error. Taubes' book also reminds us that it is in the scientific community's interest to expose spurious claims, and that, like nature, it cannot long be fooled" (Kevles 1993, 82). In the end then, science saves the day, but if this is so then why all the fuss? Why not leave cold fusion alone to disappear, as many other corrected errors in science have done in the past?

## *The Threat to Science*

Huizenga's book, along with other arguments like it, can be read in part as a reaction by scientists to a publicly perceived failure in science. As Close writes in the introduction to his book, "If these events become regarded as a norm for science then public confidence would be threatened. It is important that the public see that the test-tube fusion story is *not* typical of normal science" (Close 1991, 2). Or as Taubes writes in his conclusion, "Positive results in cold fusion were inevitably characterized by sloppy and amateurish experimental techniques. If these experiments, all hopelessly flawed, were given the credibility for which the proponents of cold fusion argued, then science itself would become an empty and meaningless endeavor. Once bad science was accepted as good enough, it could be used to prove the existence of anything, whether it existed or not" (Taubes 1993, 426). For Close and Taubes cold fusion is more than just erroneous or even bad science; it is a threat to the very legitimacy of the scientific enterprise that must be guarded against vigorously. Cold fusion is certainly not alone with respect to this worry—it is simply a very notable example—but the theme of the threat to science is echoed in a whole range of books on the popular science shelves in recent years.

Yet it is important to note that the threat to science is not just a concern of boundary workers like Taubes. Scientists working on the ground may share similar perceptions. When I asked one chemist what he thought about continued interest in cold fusion after 1990, he referred to a claim made by Pons that there would be a working CF water heater by 1993 and told me, "Remember it was supposed to take five years for this to change the world. It has been six, and the only change I have noted is a decreased respect for science and a heightened awareness of scientific misconduct."[7] Given these perceptions, the concept of pathological science might be seen as a form of ideological spin control in an attempt to reconstruct the failure of some individual scientists as a success for science as an institution. Although I do not have the space to develop this point in more detail, one can see how the reconstruction of cold fusion as pathological science in texts like Huizenga's might be linked with other public conflicts with respect to the autonomy of science occurring at the same historical moment. Particularly relevant in the early 1990's, for instance, were public perceptions of and reactions to the David Baltimore affair and cuts in federal budgets for R&D (including cuts that led to the demise of the proposed superconducting supercollider in Texas). More recently, the "science wars" can be seen to speak to this very issue (Nelkin 1996), and suddenly cold fusion and science studies are seen to be in the same boat (see Gieryn 1999).

## *Pathological Science as Boundary Work*

Rather than look at the broader context of arguments about pathological science, I will continue to focus on the specific case of cold fusion. There are two reasons for this. The first is that the thesis of pathological science has received little explicit attention in the literature on scientific controversies and boundary work, and the second is that pathological science and what I will call "pathology talk" have become, in the wake of closure, perhaps the most pervasive resource for the explanation in the scientific community of the nonexistence of cold fusion. In other words, the pathology thesis effectively competes with and then replaces other possible nuclear and chemical explanations for the phenomenon of cold fusion. In doing this, it serves to legitimate and ground claims that cold fusion is not real by offering a plausible explanation for both potential cold fusion–like anomalies in science and a continued belief in cold fusion by some scientists. Before I examine this more closely, I will glance briefly at the pathology thesis itself.

In 1953, the Nobel laureate chemist Irving Langmuir gave a lecture at General Electric's Knolls Atomic Power Laboratory on what he called "pathological science," or "the science of things that aren't so,"[8] in which he said, "These are cases where there is no dishonesty involved, but where people are tricked into false results by a lack of understanding about what human beings can do to themselves in the way of being led astray by subjective effects, wishful thinking, or threshold interactions. These are examples of pathological science. . . . There isn't anything there. There never was" (Langmuir 1989, 48). Langmuir used the term to describe a series of historical episodes in science: René Blondlot's observation of N-rays in 1903, research by Alexander Gurwitsch on mitogenetic rays in the 1920s, and the claims of extrasensory perception made by Joseph Rhine in the 1930s. All these scientists, Langmuir argued, had fooled themselves into believing in the existence of phenomena in the face of overwhelming contradictory evidence.

Drawing on these case studies, Langmuir suggested that there were six symptoms of "sick" science. A scientific claim is likely to be pathological if (1) the maximum effect observed is produced by a causative agent of barely detectable intensity, (2) the experimental observations are near the threshold of visibility, (3) there are associated claims of great accuracy, (4) there are claims of fantastic theories that are contrary to the normal experience of scientists, (5) if criticisms of the claim are met by ad hoc excuses thought up on the spur of the moment, and (6) during the course of the controversy the ratio of supporters to critics rises to near 50% and then falls gradually to oblivion (Langmuir 1989, 43–44).

In a more recent article published in *American Scientist*, Dennis Rousseau, a physical chemist at the AT&T Bell Labs, summarizes Langmuir's symptoms in terms of three basic characteristics: "The first characteristic . . . is that the effect being studied is often at the limits of detectability or has a very low statistical significance. . . . The second characteristic is a readiness to disregard prevailing ideas and theories. . . . the third identifying trait is that the investigator finds it nearly impossible to . . . conceive of and carry out a critical series of experiments. To avoid confronting the truth, the investigator selects experiments that do nothing, except perhaps add another significant figure to the result" (Rousseau 1992, 54). Rousseau uses as his case studies three of the most publicized examples of mistaken discovery in the last few decades: the claim made in the 1960s by the Russian chemists Nikolai N. Fedyakin and B. V. Deryagin to have observed a dense and viscous form of water known as polywater, the report of infinitely diluted solutions (also known as "water memory") by Jacques Benveniste at the University of Paris in 1988, and the claims of cold fusion made by Fleischmann and Pons.

Rousseau presents a range of evidence to support his interpretation of cold fusion as pathological science. To begin with, the measurements reporting nuclear effects like neutrons were barely above levels of background noise, making these claims an example of what Langmuir called "threshold observations." The important point here, which has been stressed by numerous skeptics, is that the phenomenon seems to remain at threshold levels no matter what kind of instruments are used, so that experimenters are able to claim that cold fusion is there, but just barely.

In terms of pathological science, continued belief in threshold phenomena allows researchers to plausibly discount negative observations (both their own and others'), since threshold observations are dependent on the specific experimental context—the instruments, arrangement of the apparatus, skills and experience of the experimenters, and so on. Any attempt to observe the phenomenon at a greater intensity would force the experimenters to confront the fact that the phenomenon was not really there.

Another indicator that cold fusion is pathological science is Fleischmann and Pons's insistence on proposing exotic new nuclear processes to account for their data. The claim that new physics is involved makes any attempt at verification in terms of existing theory impossible. As a consequence, negative explanations can be discounted by CF supporters as being incorrect because they are based on the faulty assumptions of the old physics.

Lastly Rousseau suggests that CF investigators have ignored definitive negative experiments. The ones he cites include experiments using light-water controls, like those of Nathan Lewis and Michael Salamon. Salamon's study is particularly valuable for the pathology thesis because it acts as a negative replication using the original apparatus (see chapter 3). This was as close as skeptics would get in terms of a public demonstration of pathological behavior similar to what R. N. Wood accomplished in the case of N-rays or James "the Amazing" Randi in the case of water memory.[9]

It should be clear from my previous arguments and from controversy studies in general that the notion of pathological science is both epistemologically and sociologically problematic. I will return to this point at the end of this chapter, but it is important to note that all of Langmuir's symptoms (and their subsequent refinements) are as open to interpretation as the experimental claims themselves and so cannot provide sturdy independent criteria for adjudicating between good and bad science. Indeed, supporters of cold fusion have no problem advancing claims that are asymptomatic. With respect to the notion of threshold observations, for instance, they point out that claims of excess heat are significant precisely because measurements are so far above threshold levels as to make conventional chemical explanations implausible. Moreover, while the first reports of neutrons and tritium were certainly barely above the limits of detectability of the instruments used, further research in 1989 and 1990 by different groups demonstrated increasing accuracy and precision. Other symptoms, such as ad hoc excuses and explanations and the manufacture of fantastic theories, are equally problematic for CF researchers, who make reference to the expertise of Fleischmann and Pons, the difficult conditions in which their results were disseminated, misinterpretations of their explanations, and substantial intellectual support from reputable electrochemists and physicists alike. As a consequence, what should be interesting for science-studies analysts is not to decide whether Langmuir and other proponents of the pathological science thesis are right or wrong, but rather to observe the effects of pathology talk on the production of scientific knowledge in the wake of the controversy.

## *Pathology Talk and Closure*

In the last chapter I argued that part of the constitution of closure in the cold fusion case involved a shift in the representation of replication by critics. By presenting negative replications as isomorphic or similar copies of Fleischmann and Pons's experiments, critics generated the conditions for participants to credibly accuse one another of experimental incompetence. Where the structural features and procedures associated with an experiment and its replication are presumed to be similar, one plausible explanation for the difference in results may be the behavior of the experimenters. At such a point controversies take on the character of what Gieryn (1999) calls "credibility contests." A consequence of this in terms of the closure of controversies is the subsequent circulation of "incompetence talk" and the polarization of participants around issues of credibility. As I argued in the last chapter, all else being equal, controversies begin to end when the statements of the actors on one side (in this case Fleischmann and Pons) are rendered less credible than statements of actors on the other.

Incompetence talk in this sense is clearly an example of boundary work, but of a different kind. As it happens, explicit accusations of incompetence amongst scientists are normally limited to informal contexts of communication (casual conversations, private e-mail, etc.) rather than rhetorical displays in public forums. Indeed, when incompetence talk leaks into the news media, it is generally frowned upon by scientists and even denied by the scientist quoted. At the end of April 1989, for instance, the *Boston Herald* published an article based on an interview with Ronald Parker, who was quoted as accusing Fleischmann and Pons of misrepresentation bordering on fraud. While the transcript of the interview indicates that this is indeed what was said, Parker denied that he was accusing Fleischmann and Pons of fraud, claiming he was trying instead to point to the dangers of their incompetence, particularly in the measurement of the gamma-ray spectrum.[10] So while incompetence talk is a form of boundary work aimed at resolving conflicts over the interpretation of experiments by undermining the credibility of the experimenters, it is a form of private or internal boundary work.

Pathology talk, on the other hand, is the more public form of incompetence talk, and I would argue that pathology talk and the thesis of pathological science became a credible explanation for cold fusion only in the wake of the controversy. During a controversy pathology talk is rare, or else it is entirely private, but as closure develops, pathology becomes a normal explanation for how the controversy occurred in the first place. Imputations of pathological science suggest a different sort of strategy at work than imputations of incompetence. Pathology supplies a general explanation for not only individual instances of incompetence but also all other experimental evidence supporting the existence of cold fusion. Thus pathological science becomes a viable theory that explains all positive replications of cold fusion. Such a strategy becomes tenable, however, only after the fact of the matter has been settled, once the controversy has ended. As long as cold fusion has been established as being not real, then all claims in support of it must be somehow in error. Note that I do not mean to suggest that closure should be understood as a discrete moment, after which pathology talk appears. What I do want to argue is that in the midst of the controversy pathology talk does not circulate easily, as opposed to after the controversy, when it does.[11] For this reason the incidence of pathology talk (but not incompetence talk) may be a good sociological indicator of closure in certain kinds of controversies.

## The Circulation and Normalization of Pathology Talk

In the first week of April 1989 (right after the press conference), as noted in chapter 2, a prominent chemist at the University of Toronto submitted a critical manuscript to a Canadian news magazine. In the manuscript, the author classified Fleischmann and Pons's claims as pathological science by comparing them to the claims of the existence of polywater (Franks 1982). The editors of the magazine, perhaps noting the general enthusiasm for cold fusion at the beginning of the controversy, declined to publish the article because it was "too harsh." At the APS meeting in May, Douglas Morrison spoke about cold fusion as a possible case of pathological science. Morrison's comments were echoed in the media, combined with an oft-cited quote from another one of the speakers, Steven Koonin: "My conclusion, based on my experience, my knowledge of nuclear physics and my intuition, is that the experiments are just wrong. And that we're suffering from the incompetence and perhaps delusion of Drs. Pons and Fleischmann."[12] In September, the director of the Brookhaven National Laboratory, N. P. Samios, and a philosopher of science, Robert Crease, published an article in the *New York Times Sunday Magazine* (Crease and Samios 1989) declaring the controversy over and marking cold fusion as just another case of pathological science, giving a list of "symptoms" loosely drawn from Langmuir's original talk.[13] In October, *Physics Today* published its reprint of Langmuir's 1953 lecture. In letters to the magazine and on Usenet newsgroups, cold fusion was compared with Langmuir's examples.

A few months later, in February 1990, *Physics World* published an article by Morrison identifying cold fusion as a case of pathological science and providing a more detailed analysis of the symptoms of cold fusion as pathology; the term has been actively used to describe cold fusion research ever since (Morrison 1990). Morrison had been active in both privately and publicly disseminating news about cold fusion via an e-mail newsletter and on Usenet, and had started deploying pathological science to explain support for CF without much resistance in July 1989.[14] While not conducting CF experiments of his own, he earned a reputation as a harsh critic of cold fusion, attending conferences long after most other skeptics and critics stopped attending.[15]

In the period between the rejected Canadian manuscript and Morrison's article, the cold fusion controversy was resolved to the extent that pathology talk became a publicly acceptable way of accounting for how the CF claims could have been made in the first place and why some people still believed them. As I argued in the last chapter, judging the acceptability of pathology talk is not a matter of making a head count, but rather of noting the circulation of such talk and judging whether there is any tangible resistance to it. In this way the nature of the phenomenon under investigation is transformed. During the summer of 1989, cold fusion as a kind of nuclear fusion began to have trouble circulating in the laboratories of would-be replicators, at conferences, and finally with funding agencies. At the same time, cold fusion as pathological science began to circulate more freely. By the end of 1990, cold fusion as pathology became more of a black-box than cold fusion as a nuclear reaction.

One of the most distressing episodes marking the conversion of cold fusion from nuclear reaction to pathology occurred in 1990 when *Science* published an article by Gary Taubes entitled "Cold Fusion Conundrum at Texas A&M" (Taubes 1990). In the article, Taubes writes of a curious situation that took place in the laboratory of John O'M. Bockris. From the beginning Bockris, Distinguished Professor of Chemistry at Texas A&M and a major figure in the field of electrochemistry, had been an influential supporter of CF research, leading a research group that was one of five at Texas A&M working on cold fusion. Bockris's group concentrated on looking for tritium. In May 1989, Nigel Packham, one of the graduate students on the team, reported higher than normal tritium concentrations in the electrolytic solution of one out of four cells. A preliminary note on these data was submitted to the *Journal of Electroanalytical Chemistry* and the group announced its results at the DOE meeting at the end of May. This was the first reported replication of anomalous tritium. In total, Bockris's group ran fifty-eight cells and found anomalous tritium in eighteen of them. The failure to measure tritium in the remaining cells was accounted for in terms of differences in the electrolysis time of some of the experiments, as well as the stocks of palladium metal used for the cathodes.

Bockris had produced important confirmatory data for the fusion hypothesis, but there was also significant criticism. The major concern was contamination. Had Bockris taken precautions to prevent tritium contamination from outside sources—either tritium that already existed in the palladium cathode or tritium that had found its way into the electrolytic solution during the preparation of the cell? According to Bockris, great care was taken in the preparation of the cells, and few would doubt his expertise in such matters. The palladium had been purified through electrolytic separation and then melted at 1,534 degrees Celsius to produce the cathode wire. Bockris argued that this process would effectively remove or destroy any existing tritium ions. The tritium measurements were accurate.

A local controversy of sorts then developed between Bockris and Kevin Wolf, a chemist who was leading another search for tritium at Texas A&M. Wolf claimed to have not detected any anomalous tritium in his own cells, and he further argued that in preparing his cathodes in a manner similar to that of Bockris he had found significant levels of tritium contamination. To compound the conflict, Wolf had even tested some of Bockris's cells. Bockris responded that Wolf's cathodes were not the same as his and that Wolf's calculations were in error.

The debate might have continued in this manner for some time, but it started to take a different turn once Taubes became involved. During the summer of 1989 and into the fall, Taubes had maintained a relationship with Bockris's lab; he had access to lab notes and conducted several interviews over an extended period. Drawing on comments from Wolf and Charles Martin, Taubes suggested the possibility that the anomalous tritium was the result not of contamination but rather of the deliberate spiking of the cells with tritiated water. He supported the hypothesis by suggesting that Bockris's tritium measurements were very high, sporadic, and seemingly not reproducible. Since the majority of Bockris's cells generated no significant results, it seemed unlikely that there could not be any kind of systematic error. Whatever was happening was either random or planned. Taubes noted that Bockris's tritium results coincided with visits from members of the Electric Power Research Institute (EPRI). Perhaps someone was spiking the cells with tritium to generate significant results in order to attract more funding.[16]

In this episode we see again the potential consequences of similarity arguments in debates over replicability. If the experimental protocols for a positive replication (Bockris) and negative replication (Wolf) are perceived to be the same, it can be argued that the difference lies with the competence of the experimenters; Bockris did not prepare his electrodes carefully enough, Wolf did not consider all the data, and so forth. While incompetence talk is not unusual during controversies, it is only in the closure of the debate that incompetence as pathology becomes a more publicly plausible claim. In June 1990, *Science* published Taubes's article, and an administrative investigation into the cold fusion research and allegations of fraud began at Texas A&M.

As it happened, the Texas A&M review panel distanced itself from the fraud hypothesis and supported the ongoing efforts of Bockris and his research group. The final report of the investigation, released in October, helped to clear Bockris and especially his graduate student, Nigel Packham, of any wrongdoing.[17] The scientists on the panel resisted the claim that cold fusion research was "bad science," calling instead for better experiments while citing issues that were contested during the summer of 1989. But although resistance to the fraud thesis appears in the Texas A&M report and in numerous letters sent to the editor of *Science* after the publication of Taubes's article, the lack of any formal retraction from the journal, followed by the publication of Taubes's book in 1993, helped to circulate the idea of cold fusion as a kind of pathological science. The effects of this can be seen, for instance, in a literature review of scientific deception published by the British Library, where cold fusion is included as a case of widespread self-deception in science and Taubes's book is listed as the primary citation (Grayson 1995, 32).

The normalization of cold fusion as pathology took an even more curious turn in Italy. In October 1991, the Italian newspaper *La Repubblica* published an article by its science editor, Giovanni Maria Pace, that was ostensibly a review of Alexander Kohn's *False Prophets: Fraud and Error in Science and Medicine*. Originally published in English in 1988 and published in Italian in 1991, Kohn's book revolved around several case studies of "fraud," "gross error," and "pathological science." Pace, drawing on arguments from the book, accused Fleischmann and Pons and Italian CF researchers of being "false prophets" and of perpetrating a fraud on the scientific community and the public. His accusation was extreme to say the least. Pace argued that cold fusion was not real because there was no experimental evidence for neutrons (a "necessary" consequence of fusion phenomena), that there was a massive majority of failed replications versus successful ones, and that scientific interest in cold fusion was fueled by its economic implications and not by sound science. Further, Fleischmann and Pons had contravened normal scientific practice by announcing their results at a press conference, and there was evidence that they had withheld and distorted their experimental data.[18] Finally, in an interesting expression of boundary work, Pace wrote, "The crime of 'damaged truth' is outrageous and it is perceived as a scandal by even the staunchest pragmatists because it has been committed in the very temple of truth. If in the scale of human abjection there is a being that is to be located even below the fornicating priest, or the pedophile teacher, this is the lying scientist."[19] Fleischmann and Pons submitted a letter to the editor arguing for the legitimacy of their claims, and part of this was published on November 6, 1991, under the title "We Are Not False Prophets." Published along with their letter was a reply from Pace, under the title "Yes You Are."

After submitting further letters of correction, which were not published, Fleischmann and Pons, along with the Italian researchers Guiliano Preparata, Tullio Bressani, and Emilio Del Giudice, decided to sue *La Repubblica.* The plaintiffs' position was based on the argument, set out in Fleischmann and Pons's letter, that they had done nothing either immoral or unscientific. Their claims had in fact been replicated in both the United States and Italy. They had not committed fraud or misrepresented their data; moreover, a large number of scientists around the world believed that the scientific work on cold fusion was sound and worthwhile. The plaintiffs asked that published articles on cold fusion be admitted as testimony of the legitimacy of Fleischmann and Pons's scientific work.

Expert testimony in the form of detailed reports on the scientific data was elicited by the court, the plaintiffs, and the defense. (Each party employed its own experts.) The review of the written testimony in the final court transcript briefly rehearses most of the arguments I presented in chapter 3 and so becomes an interesting context in which the issue of interpretive flexibility is played out. The conclusion of the judge was as follows:

> On several occasions Martin Fleischmann and Stanley Pons provided contradictory data, omitted the contribution of Jones to these studies, issued declarations to the media on the developments in their research which were absolutely devoid of any reference to reality. . . . All these factors make us very perplexed as to the behavioral correctness of these two scientists. . . . It must therefore be held that the expressions of severe criticism used in the articles by Giovanni Maria Pace with regard to the supporters of cold fusion are justified by the existence of a substantial disagreement and questioning on the part of the scientific community. Not only with regard to attack on the theoretical status of this research but also with regard to the way in which the relevant data have been published and the conclusions as to future developments have been taken. . . . There is a further confirmation of the appropriateness of the critiques measured against Fleischmann and Pons which is that even considering the time elapsed from the beginning to the present proceedings no further

> steps have been taken in research in so-called cold fusion and in fact the scientific community is in the process of becoming disinterested in such an issue. Therefore this court believes that the sentences used in the article published . . . represent an expression of the right of reporting and therefore they are not defamatory.[20]

The decision of the court supports not just Pace's right to free speech but also the basic content of Pace's claims by speaking directly to the evidence for cold fusion. It conflates a lack of "behavioral correctness" with the perceived lack of experimental data so that the former becomes a cause of the latter. That is, the scientists' desire for cold fusion to be real combined with the absence of data prompts the impropriety noted by Pace and defended by the judge. Without evidence of cold fusion, Fleischmann and Pons have no excuse for acting in the ways they are accused of. In the courtroom there is no sense of any ongoing controversy over cold fusion: there is no new evidence, and fewer and fewer scientists are interested in the issue (despite the claims of the defense). For this judge, the case is literally closed, and Pace's accusation of "false prophesy" becomes not just the normal explanation for cold fusion found in the media, but an explanation supported by the Italian courts.

## *Pathological Science and Scientific Work*

The normalization of pathology talk in the context of Taubes's book and the Italian court judgment shows that the reality of cold fusion is no longer the issue. Instead, the focus is on the characteristics of cold fusion as pathology. Interest shifts from debates over the presence or absence of neutrons to the various symptoms of pathological science, along with concern over how to warn scientists against the risks of self-deception and protect the public from being taken in. Is this to say then that cold fusion is no longer an object of natural science? That may be a fair argument. Perhaps in focusing on popular books and courtrooms I have not been attentive enough to the attitudes of practicing scientists. Morrison, Huizenga, and Close are certainly scientists, however, and favorable reviews of their writing were published in respected science journals, and scientists were consulted as expert witnesses in the Italian court case. But still, perhaps when scientists are not in their laboratories or writing technical papers, they are not really doing science. Morrison, Huizenga, and Close didn't actually perform any cold fusion experiments; they just wrote about others' experiments for audiences of nonscientists.

I thus seem to have traced a shift in the contexts in which cold fusion has been produced. It has moved out of the laboratory and into the world (Latour 1983), but in a way that severs the relationship between the laboratory and the world and leaves us with the impression that boundary work is merely politics that happens outside or around scientific work. Indeed, this is the impression generated by the Texas A&M review panel's final report: "In cold fusion research attempts to establish scientific priority or to counteract criticisms appear to have resulted in some breakdown of scientific objectivity. . . . Based upon our review, the committee concludes that the problem of proving or disproving the existence of cold fusion remains an experimental one which will only be settled by appropriate and well-executed experiments. The local programs appear to have settled into a normal and well-planned mode of operation for scientific research and one can hope that they will contribute to understanding the phenomenon."[21]

This is not quite the impression I wish to generate. The pathologization of cold fusion is not simply a "breakdown of scientific objectivity," nor is it just a political exercise in demonstrating the authority of science; it also bears directly on the execution and interpretation of experiments as part of the management and organization of scientific knowledge and practice. A useful way of thinking about this is to describe cold fusion as a kind of practical limit on what counts as scientific knowledge. Pathology talk constitutes cold fusion as a taboo object for scientists while providing a means for evaluating similar observations and claims that seem similar to cold fusion.

In the wake of the controversy, the pathology talk of boundary workers like Huizenga and Close has been picked up and deployed by other participants. When Charles Martin (see chapter 3) submitted the final report on his cold fusion experiments at Texas A&M to the chair of the chemistry department in February 1990, he wrote, "None of the 83 cells which were run by my students in my laboratory have produced tritium levels above those predicted by the separation factor. . . . the report of Pons, Fleischmann and Hawkins is incorrect. We believe that cold fusion, as it is currently defined, does not exist."[22] Along with his report, he submitted a copy of Langmuir's lecture as well as Morrison's just-published article "The Rise and Decline of Cold Fusion."

In the transcript of a roundtable discussion presented in the newsletter of the Lawrence Livermore National Laboratory, one ex–cold fusion researcher, Keith Thomassen, gives his summary of the cold fusion episode: "I suspect this matter will largely subside in time, and we will have one more anecdote on how good science should be done and how some science is not done well. Scientists may be reminded to be a little more cautious. If instances of pathological science serve as periodic refreshers of the scientific method, then they probably have some value in that sense" (Lawrence Livermore National Laboratory 1990, 23). Another ex-CF researcher, David Williams, also deploys pathology talk in a review of Huizenga's book for *Physics Today*: "After reading the book, I asked myself what has changed with respect to cold fusion since it was written. The answer must be, not much. Public presentations of the topic tend to fudge experimental details, give only partial results or results taken out of context and adduce relationships between unrelated experiments while glossing over inconsistencies. It seems to me that in their desperate need to attribute phenomena to the existence of a nuclear effect, the proponents of cold fusion have failed to distinguish in their results what is real from what is imaginary" (Williams 1993). These comments illustrate the ways in which scientists may deploy public pathology talk as a means of making sense of their own experience with cold fusion in the laboratory.

Texts like Huizenga's are not only apologias for science; they are also warnings to scientists to avoid cold fusion or face becoming pathologized. This may seem to be a sinister way of putting it, but the transition in the configuration of cold fusion from science to pathology can be quite explicit. As Huizenga writes, "Room temperature nuclear fusion without commensurate amounts of fusion products is a delusion and qualifies as pathological science, defined as the 'science of things that aren't so'" (Huizenga 1992, 204). This kind of transition is not simply rhetorical posturing. Although Bockris and his group were cleared of fraud charges at Texas A&M, the normalization of pathology talk in the case of cold fusion took its toll when it came time for Nigel Packham to do his Ph.D. defense. In August 1990, Packham went through a harrowing experience dealing with one committee member who sternly objected to material drawn from cold fusion research in his thesis. While Packham's thesis started as a study of the production of hydrogen in bacteria, he had shifted to work on tritium measurements in cold fusion cells. For this committee member, Packham's tritium measurements constituted sloppy science, and although a deal was struck whereby Packham would provide extended written responses to the questions of the dissenting committee member, the thesis was not accepted. Finally in 1992, Packham handed in an acceptable thesis. He had been forced to rewrite it, excluding all the material on cold fusion.[23]

In the case of Packham's Ph.D., cold fusion and research that supports it are ultimately not defendable, and while the conflict of personalities did play an important part in the Texas A&M case, the problem extends to other institutions, where cold fusion is not considered a suitable topic for students. As one physicist told me, "I had two or three students from the physics department come by and work with me on it [cold fusion], and one of them wanted to do a physics department bachelor's thesis on it. And the physics department failed to approve that, they told him go find somebody else to work with."[24] Here processes of delegitimation and pathologization are tied directly to the issue of labor. Students are advised not to work on cold fusion because it could jeopardize their careers.

Before the cold fusion controversy, fusion without commensurate fusion products was considered an improbability given the existing state of knowledge. The cold fusion claims challenged this idea of conventional commensurability and opened the door for the legitimate study of alternative fusion physics. Cold fusion appeared to scientists as an opportunity, but in 1990 it became more of a liability. The pathologization of cold fusion attempts to shut down this alternative physics not because it is improbable but because it is delusional. As a result of the controversy, scientists now have a way of classifying experimental effects and theoretical claims that are similar to those of cold fusion.

Both before and after the cold fusion controversy, there were numerous claims of alternative physics for generating useful energy, and while most of these claims have been treated with a great deal more skepticism, they have become strange bedfellows for cold fusion. As I shall argue in chapter 6, the association of these "free" or "new" energy claims with cold fusion research may even help them gain more legitimacy than they might otherwise have (at least amongst networks of cold fusion researchers). On the other side, however, certain other lines of scientific investigation that were perfectly legitimate before 1989 became more problematic in the aftermath of the cold fusion controversy. Investigations into piezonuclear fusion, for instance, which looked at the possibility of nuclear reactions in the Earth's mantle by testing for tritium in volcanoes, had generated promising but inconclusive results. After 1990, funding for this research became more difficult to obtain.

Another area of investigation is sonoluminescence, in which sound energy is used to excite bubbles of liquid, resulting in the emission of light energy. From studies of energy emissions in experiments, some scientists speculate that the temperature inside the bubbles might be high enough to produce fusion reactions—a process referred to as sonofusion. While sonoluminescence is an unproblematic experimental effect, sonofusion is controversial, and proponents of this research have tried to distance themselves from the hyperbolic claims of cold fusion by remarking that if sonofusion is indeed possible, it is unlikely to produce useful energy.

These examples show that while boundary work serves to enforce the cognitive authority of science in the wider culture, it is also an important means of stabilizing scientific knowledge. Pathology talk helps to reinstantiate the conventions of nuclear physics by defining fusion as a process that produces "commensurate amounts" of heat energy and nuclear ash. Fusion reactions that do not produce energy and ash become cold fusion–like, and with the translation of cold fusion from nuclear reaction to pathology, other cold fusion—like effects come to be seen as mistakes or self-delusion. Further, pathology talk continues to "mark" (Hess 1992) any researcher that supports cold fusion–like claims as being a sloppy or misguided scientist.

## *Beyond the Core Set*

The perceptions of scientists who were not participants in the controversy is harder to gauge, but the sense that pathology talk has entered more general circulation is supported by incidental comments made by scientists to whom I tried to describe my research for this book. Comments like, "If you want to know how *science* works, then why are you studying cold fusion?"[25] are indicative of a general perception amongst scientists that cold fusion is not only not real, but is pathological science. Some science-studies scholars, as I have already indicated, share this perception as well; not only is cold fusion pathological science, but this book may be an instance of pathological-science studies by extension. As one sociologist told me in an e-mail discussion about my study of cold fusion research, "If we are interested in the interaction of/ role of/sociology of science in society then it makes sense to define science as it is defined by the society in which it resides. Therefore activity that is considered completely unscientific does not fall within the scope of social studies of science."[26]

Beyond this, cold fusion has become a popular term used by scientists talking to the media about any outrageous claim lacking evidential support. In a *New York Times* article, for instance, critics of the neurologist Stanley Prusiner referred to the prion hypothesis in virology as the "cold fusion of infectious diseases" (Kolata 1994). Cold fusion as pathology can also be traced farther as it appears in lectures and courses on ethics in science,[27] in the pages of the satiric *Journal of Irreproducible Results*, as jokes in television sitcoms (*Murphy Brown* and *The Simpsons*), and even in popular advertising. Cold fusion has also been adopted as the product names for a kind of web application software and for protein bars.[28]

One poignant example of this is a full-page advertisement for Pepsi-Max that appeared in the *Toronto Globe and Mail* on January 27, 1995. The ad features a bottle of the beverage against a bright yellow background. At the top of the ad in large black letters are the words "COLD FUSION," and the caption at the bottom of the ad reads, "Pepsi just did what many would have thought impossible. We've combined all the full, rich taste of a regular cola with only one third the calories of regular Pepsi, to give you a revolutionary new cola that delivers the best of both worlds. Proof that nothing is impossible after all." The ad indirectly references the state of cold fusion in 1994—an impossible revolution. Cold fusion is deployed as not just a revolution that didn't happen, but a revolution that couldn't have happened. In 1994 the historical circle is complete, and cold fusion becomes an impossibility from the beginning. Here cold fusion is made to belong to the realm of fanciful belief, UFOs and fairies, N-rays and orgone energy—and ghosts. That Pepsi can do with advertising copy what scientists could not do in the laboratory only reinforces the translation of cold fusion from science to not-science.[29]

This brings us back to *The Saint*; what is configured as "delusion" amongst scientists has become "imagination" in contemporary popular culture. Another film, made prior to *The Saint*, articulates the distinction between science and cold fusion very well. In the 1994 film *I.Q.*, Albert Einstein develops the idea of cold fusion as a ruse to help a garage mechanic friend impress his brainy niece. The mechanic takes credit for the discovery, becomes a hero, and wins the girl. At the end of the film, a jocular Einstein (played by Walter Matthau) reveals the discovery as a fake perpetrated in the name of love. The film becomes all the more interesting in that Keith Johnson, a chemist and cold fusion researcher working at MIT, wrote the original screenplay. Johnson told me that his script was completely revised for the film and that the love interest in his script developed around the "real" discovery of cold fusion, not a ruse. In a move that parallels the postclosure survival of cold fusion research itself, Johnson has circumvented Hollywood and independently produced a film called *Breaking Symmetry* based on his original script.

While cold fusion is portrayed as "real" in *The Saint*, the sense that its reality is fantasy is enhanced by the gendered portrayal of its discoverer. Emma Russell is conspicuous in her unscientific behavior. She is scatterbrained, emotional, and irrational. She is never shown doing research in a lab, but instead arrives at the solution to the cold fusion problem through intuition or spontaneous flashes of insight. Emma's behavior is reminiscent of boundary workers' public representations of Fleischmann and Pons as being irrational and unstable and as being unable to supply adequate justifications for their claims. Fleischmann and Pons's idea, like Emma Russell's, is portrayed as a flight of fancy. In the film, Russell's behavior is rewarded (she gets fame and a heteronormative relationship), while in reality Fleischmann and Pons's behavior is the subject of ridicule.

In science and in popular culture, cold fusion takes on the attributes of delusion and fantasy—the very opposite of reality. Note the following comment from James "the Amazing" Randi, a professional magician who specializes in debunking "pseudoscientific" claims: "At the risk of being unbearably realistic, I must tell you that Elvis Presley is really dead, the sky is not falling, perpetual motion is a chimera, cold fusion is a dead duck, the earth is not flat, and the fault lies not in our stars, but in ourselves" (Randi 1994). In the wake of the controversy, cold fusion has not disappeared at all; it circulates in Randi's statement as a kind of object alongside other nonexistent objects, including Elvis, a falling sky, perpetual motion, and a flat Earth. This class is made up not of fictional things (like mythological beasts, or characters in a novel) but of mistaken beliefs—beliefs that were or are thought to refer to something real but do not in fact do so. What is important here is that this class of mistaken beliefs is not absent from the history of science (they are not represented by a silence) but is deployed by various actors as means of establishing the boundaries of legitimate practice and the stability of conventional knowledge.

## *Voodoo Science and Science Studies*

More than ten years after the initial press conference, cold fusion still circulates as pathological science. It appeared recently in a book entitled *Voodoo Science*, by Robert Park (2000). Park was trained as a physicist, with an expertise in surface physics. While he did not perform any CF experiments in 1989, he became well acquainted with the controversy in his capacity as the director of the Office of Public Affairs of the American Physical Society in Washington, D.C. For Park, cold fusion belongs with other cases of outlandish claims pretending to be science.[30] As he explains in the preface to his book,

> As I sought to make a case for science, however, I kept bumping up against scientific ideas and claims that are totally, indisputably, extravagantly wrong, but which nevertheless attract a large following of passionate, and sometimes powerful, proponents. I came to realize that many people choose scientific beliefs the same way they choose to be Methodists, or Democrats, or Chicago Cubs fans. They judge science by how well it agrees with the way they want the world to be. A best-selling health guru insists that his brand of spiritual healing is firmly grounded in quantum theory; half the population believes the earth is being visited by space aliens who have mastered faster-than-light travel; and educated people wear magnets in their shoes to restore natural energy. Did we set people up for this? In our eagerness to share the excitement of discovery, have scientists conveyed the message that the universe is so strange that anything is possible? What can we tell people that will help them to judge which claims are science and which are voodoo? (Park 2000, viii–ix)

Park's recounting of the story of cold fusion repeats a number of the criticisms from 1989 (improper controls, inconsistent heat and neutron measurements, tritium contamination, etc.), but his account does not address more recent work dealing with those criticisms. In the ten years between the press conference and the publication of Park's book, Fleischmann and Pons published a number of other papers, and these were accompanied by scores of papers from other scientists claiming not just confirmation of cold fusion but "better" confirmation. Yet for Park the scientific investigation into the phenomenon ended in 1989–1990; what is left is pathological science. The continuing work of cold fusion researchers presents a problem for Park: "Although cold fusion disappeared from the front pages years ago, a dwindling band of believers has gathered each year since 1989 for a meeting at some swank international resort to share the results of their efforts to resuscitate cold fusion. . . . These are scientists; they are presumably trained to view new claims with skepticism. What keeps them coming back each year with hope in their breasts? Why does this little band so fervently believe in something the rest of the scientific community rejected as fantasy years earlier?" (Park 2000, 12–14).

Park is hard pressed to provide an answer to these questions. It is one thing for the public to be fooled by voodoo science and quite another for scientists to be taken in a large scale. Park's solution is to minimize the scale and significance of cold fusion research after 1990 and imply that scientists who still believe cold fusion is real suffer from the same pathology as Fleischmann and Pons: "They were trapped by the enormity of their claim. They could not now acknowledge that there had been no fusion without admitting that they had previously exaggerated or fabricated their evidence. They had become world figures, so they now stood to be disgraced before the entire world. What began as wishful interpretations of sloppy and incomplete experiments had evolved into deliberate obfuscation and suppression of data" (Park 2000, 122–123). Park's account, like Huizenga's, is little more than ideological boundary work. In addition to promoting the ongoing pathologization of cold fusion in a regular e-mail newsletter, Park has been active in barring cold fusion sessions from APS meetings and preventing federal funding for research.

I have no trouble confessing that I consider Park's account of the cold fusion case to be extremely problematic, and it is worth noting the methodological difficulties in making such a confession. I have been arguing that boundary work and its public manifestation in pathology talk constitute part of the means that make science work. For better or for worse, Park's text is part of what it means for sociologists to understand science in action, and it plays an important role in shoring up conventional knowledge and practice with respect to nuclear physics. This is a strong social-epistemological claim. Nuclear physics becomes a more stable and reliable body of knowledge with the circulation of cold fusion as pathology. Not only is this useful to understand, but stable and reliable knowledge about the world is also a good thing for societies and cultures to have.

What troubles me about Park's account is that with the pathological science thesis we are drawn into a kind of social psychology of scientific error. This is no longer about the experimenter's regress and debates over what counts as a successful replication. Park's account of the behavior of Fleischmann and Pons and other cold fusion researchers draws upon social causes and psychological motivations (greed, wishful thinking, peer pressure, celebrity, personality politics, and egotism) as a means of explaining how scientists can fool themselves. Indeed, Park's account echoes that of other boundary workers like Frank Close: "I don't really want to talk about cold fusion—what's been going on for the last five years because that would take too few minutes [laughter]. What I wanted to do was to tell you a story of intrigue and psychology, about what can happen when you convince yourself that something is there to be found and when the evidence shows that it isn't, you reinterpret the evidence to fit in with your preconceptions."[31]

Should sociologists and psychologists take offence at this? Should someone without appropriate training in sociology or psychology be permitted to comment authoritatively on such matters? I am less concerned about these questions than I am with the idea that, under the right circumstances, the discussion of pathological science may be an appropriate place for science-studies analysts to speak up about scientific controversies. On such occasions it may be useful for science-studies scholars to temporarily abandon any pretensions to neutrality since we lay claim to a better sociological account of what is going on than the thesis of pathological science can provide. In configuring cold fusion as pathological science, Park and others open the door to alternative sociological accounts, including symmetrical explanations that do not see truth in science as any less amenable to sociology than error. Despite the science wars, perhaps science needs sociology after all.

In the previous chapter, I discussed the ways in which the debates over the existence of cold fusion came to a close. In this chapter I have described the ways in which that closure has been maintained, as a means of both forestalling a potential reopening of the controversy and further stabilizing knowledge about the physical world. This is accomplished through practices that construct cold fusion as a pathological object. It is in the wake of closure extended by the circulation of cold fusion as pathological science that CF researchers continue their work today. What follows in the next two chapters is a different sociological view of cold fusion research after 1990—not as pathological science but as undead science.

# Chapter 5 The Afterlife of Cold Fusion

## *Heat-after-Death*

In October 1993, more than four years after their initial press conference, Fleischmann and Pons reported a new phenomenon associated with their cold fusion experiments. They called the new effect "heat-after-death." (Fleischmann and Pons 1993, 1994a). The effect can be observed under the right conditions when the electrolytic solution is driven to the boiling point and allowed to boil dry. In the absence of electrolyte, the circuit breaks and the electric current stops moving through the cell, so that energy is no longer being added to the system. Despite this absence of applied energy, Fleischmann and Pons claimed to have observed high levels of excess heat being generated in their apparatus for a number of hours. If heat was being generated in the complete absence of energy input, where could it be coming from if not from a reaction within the cell? Even if the excess heat was not due to nuclear fusion, at the very least these new observations should prove that the excess heat effect is real and not an artifact of external heating or a recombination of gases, as many critics had previously argued.[1]

This time, Fleischmann and Pons's paper had been peer reviewed and accepted for publication in *Physics Letters A*, a fairly prestigious journal. There was no press conference and no attendant hyperbole, although the article did prompt a few reports in the media,[2] a critical response from Morrison (1994), and a brief flurry of messages in the Usenet newsgroup sci.physics.fusion. What was most notable, however, was the general lack of interest of the majority of scientists who had participated in the controversy and sparred with Fleischmann and Pons in 1989–1990. Fleischmann and Pons had produced new evidence to help bolster their claims and allay criticism. They chose to publish in *Physics Letters* A for this reason, but the reengagement they had hoped for did not arise. Far worse, a few scientists expressed their private opinion to the editors of *Physics Letters* A that the publication of Fleischmann and Pons's paper had tarnished the journal's reputation.[3]

Yet it would be wrong to suggest that Fleischmann and Pons were simply crying wolf and no one was listening. Quite the contrary: Fleischmann and Pons still had an audience, but it was not as easy to find as in 1989. In December 1994, they presented their heat-after-death results at the fourth International Conference on Cold Fusion (ICCF–4), held at a resort hotel on the Hawaiian island of Maui. With funding from the Electric Power Research Institute and organizational support from the Stanford Research Institute, conference attendance for that year was up from the previous cold fusion meetings. Over two hundred participants came from several countries, and there were over forty scientific papers given in sessions on theory, materials, calorimetry, and nuclear measurement. Nearly all of the papers presented evidence confirming the anomalousness of cold fusion, although few offered any kind of systematic explanation for the measured effects. For the scientists at the conference, most of whom had been working on CF since 1989, Fleischmann and Pons's new observations were treated as an important development for the field. Could heat-after-death mean renewed life for cold fusion?

Heat-after-death certainly seemed like a potent demonstration of anomalous heat. As in a magician's levitation trick, where the table supporting a prone woman is removed, leaving her floating in midair, Fleischmann and Pons removed the input energy, and the temperature of the cell did not decrease as it should have. They had left the excess-heat effect but removed the source of an artifact.[4] This was certainly impressive, but what scientists wanted and needed to know was how to pull it off for themselves. What heat-after-death did not do, it seemed, was supply eager (and in some cases desperate) CF experimenters with any greater sense of how to reproduce the effect consistently on their own.[5]

Since 1989, CF experiments had been plagued with the problem of inconsistent reproducibility. While some researchers claimed near total reproducibility under specific conditions, many others who had measured CF effects more than once were having trouble reproducing the effects consistently. Critics cited this inconsistency as justification for the belief that cold fusion was not reproducible and that the sporadically observed effects were the result of errors or artifacts, but a significant number of CF researchers disagreed. Although the effects appear sporadically, that does not mean that the phenomenon is not real, nor does it preclude the possibility that innovations in the experimental apparatus may produce the effects more reliably. What was unclear was whether sporadic and variable effects were a property of the phenomenon itself (in which case commercial energy generation would probably not be a viable option in the end) or whether a protocol could be devised that might produce the anomalous effects more reliably.

Fleischmann and Pons's heat-after-death experiments did suggest that the phenomenon was more stable in cells running at higher temperatures, but this came at the expense of more complex calorimetry that was beyond the skills and resources of many of the other researchers at the Maui conference. In the end there would be no breakthrough for cold fusion in 1993, and today reproducibility continues to be a problematic issue. But in spite of this, heat-after-death was doing something else important in Maui. In the wake of the closure of the controversy and the pathologization of cold fusion, heat-after-death was helping to establish solidarity amongst researchers, constituting a collective identity by providing a rhetorical affirmation of the legitimacy of cold fusion research along with a forceful rebuttal to the criticisms of skeptics. As one conference participant commented in an electronic discussion of the heat-after-death paper, "It proves their point exactly. It is an elegant paper, describing an elegant and wonderful experiment. It demonstrates beyond any rational doubt that the heat from CF is beyond chemistry."[6] In the context of the cold fusion conference, Fleischmann and Pons's new results represented both progress and hope for researchers who were feeling under siege, and this helped bring the researchers together as a group. In 1993 at least, heat-after-death provided confirmation of the sense that researchers were at least on the right track.

## *The Science of Cold Fusion*

This chapter and the next form the empirical basis for the argument that contemporary cold fusion research can best be understood as undead science: an organization of scientific culture and practice that takes its shape from being on the losing side of a controversy. "Losing" in controversies is a historical consequence brought about by shifts in the social and material organization of scientific practice. These shifts result in the dynamics of closure (of the kind discussed in chapters 3 and 4) that produce the truth or falsehood of scientific claims. The cold fusion controversy did not end because Fleischmann and Pons were wrong; it ended because CF researchers found themselves lacking social and material resources to argue that they were right. Yet despite this, cold fusion research has continued after the generally acknowledged end of the controversy in 1990, so resources must be coming from somewhere. How

CF researchers are able to do their work, what kinds of research are occurring, and how the research is collectively organized are the questions that motivate the argument of the next two chapters.

In the first instance I am less concerned with making a case for how post-closure research happens than I am with making the case that it happens at all. As I argued in chapter 4, cold fusion as pathological science has become the normal explanation for post-closure CF research. While this is an effective means of reproducing closure and reinforcing the epistemic boundaries of conventional nuclear physics, the prevalence of pathology talk has the effect of publicly eliding or even suppressing a collective practice that is no less "scientific" than any other kind of mainstream science. Because we live in a world in which cold fusion research is generally perceived as being pathological science, it will require a little extra work on my part to demonstrate that there is nothing particularly pathological going on in the laboratories of cold fusion researchers while at the same time arguing that CF research is not normal science.

Let us begin by considering one view of what we might encounter at the end of a controversy:

> When a scientific controversy reaches closure, there are winners and losers. If new claims are rejected outright by the majority, the winners write their "told-you-so" books and papers and go back to their previous scientific lives; for them there is a resumption of business as usual. As for the losers, they may disappear too, or they may form a "rejected science"—a determined rear-guard action that the mainstream cannot or will not understand. Another possibility is that a new post-closure science develops that is a modification of the science at the center of the core-set controversy. When this happens, the mavericks are expelled and a "core-group" takes over. (Collins 2000, 824–825)

Thus Collins sets the stage for a discussion of the survival of the rejected science of high-visibility gravitational radiation. Of particular interest here is the tension between a core group, defined as a socially solidaristic group working legitimately within the normal institutional frameworks of science, and a rejected science, defined as "a collection of scientists obstinately refusing to give up their ideas in spite of the crushing consensus that surrounds them" (Collins 2000, 825). In the case of gravity wave research, the controversy over Joseph Weber's claims to have detected high-visibility gravitational radiation in the 1970s ended with the formation of a legitimate core group focused on low-visibility gravity waves and a rejected science obsessed with pressing the high-visibility claims. While the social relations of the core group tend to be solidaristic and consensual, relations amongst members of the rejected science can be characterized by loose associations and plurality with respect to belief and practice.

As Collins (2000) demonstrates, the gravity wave case is immediately comparable to the case of cold fusion (amongst others), and pursuing this comparison sets the stage nicely for making sense of CF research as undead science. The problem is that CF researchers display the social properties of both core groups and rejected sciences. CF research is characterized by a core group–like degree of solidarity and shared understanding, but all CF researchers experience the stigma of rejected science. At the moment there is no appreciable legitimate middle ground for CF researchers, and they remain caught between establishing a stable core group within science on the one hand and suffering permanent rejection on the other.

It is important to point out that when working with the categories of core group, rejected science, and undead science we are not trying to make distinctions based on who is really doing science.[7] On the most nominalistic level, post-closure CF research is science because CF researchers "collectively use the tools and procedures of science within the institutions of science" (Collins 2000, 825). Collins makes this case with respect to the rejected science of high-visibility gravitational radiation, but the observation also applies to cold fusion. CF researchers work in university physics, chemistry, and engineering departments, they are well trained, and many have reputations as outstanding scientists for work they have done in other areas. For the most part, they value the peer-review process and publish with some success in the same journals as their mainstream colleagues. They do experiments following methodological norms that are indistinguishable from those of other experimental sciences; they value precision and accuracy in measurement and the reproducibility and robustness of experimental effects, even if they disagree as to what the effects are. They seek to improve their experiments both individually and collectively, and they seek out audiences and forums where their work can be discussed and refined by peers. To the casual observer there is nothing in the daily routine of most CF researchers that would indicate that what they do is in any way unscientific. Given all this, we can follow Collins in arguing, "To conclude that, on the basis of their observable activities, this group is doing something other than pursuing a particular strand of science would be to make a judgment *within* natural science, and this is not the business of the sociologist" (Collins 2000, 825). With this point in mind we can begin to look at what cold fusion research as a particular strand of science consists of.

## *The International Conferences on Cold Fusion*

Perhaps the most tangible evidence of an organized post-closure scientific life for cold fusion is a regular series of meetings attended by CF researchers from around the world. The meetings are known as the International Conferences on Cold Fusion, and there has been one nearly each year since 1990.[8] ICCF–1 was held in Salt Lake City on the anniversary of Fleischmann and Pons's initial press conference. The second conference was held at Lake Como, Italy, in late June 1991, and the third was held in Nagoya, Japan (1992). These were followed by ICCF–4 on Maui, Hawaii (1993), ICCF–5 in Monte Carlo (1995), ICCF–6 in Hokkaido (1996), ICCF–7 in Vancouver (1997), ICCF–8 in Lerici, Italy (2000), and ICCF–9 in Beijing (2001). Attendance at the conferences has ranged from two hundred to three hundred fifty participants, with the majority of researchers coming from the United States, Japan, Italy, France, and Russia.

Professional meetings can be seen as part of the normal life of any scientific field. In an institutional context, scientists are often expected and encouraged to present their research at meetings, and at universities conference papers may figure into the determination of relative merit amongst the faculty. Yet irrespective of the institutional incentive for scientists to present papers, conferences are deemed to be of value by scientists as means of networking, learning about the state of research in a field, and gaining exposure to new ideas. During scientific controversies, conferences become even more valuable. Throughout the summer and fall of 1989, the many cold fusion sessions that were sponsored by scientific organizations were crucial moments for participants to come into direct contact with one another, exchange information, and formulate decisions as to how to proceed. The American Physical Society meeting in May 1989 figured prominently, for instance, in many scientists' decisions to stop their replication attempts. At other meetings, however, after having met with experimenters who had been successful, some scientists decided CF research was worth continuing.

Given the centrality of conferences and meetings to scientific life, it is surprising that there have been few specific sociological studies of the role of conferences in the production of contemporary scientific knowledge. However, we can certainly infer from work in the sociology of organizations and communication that conferences are significant in that they facilitate both deeper and broader social ties (and consequently greater social cohesion and integration).[9] This is accomplished through direct face-to-face interaction, as opposed to routine indirect and mediated forms of communication like journals, letters, phone calls, or electronic mail.

There are two consequences of this relevant to work in science studies and cold fusion. First, conference interaction facilitates the amelioration of conditions of experimenter's regress by allowing for greater transfer of the context-specific issues associated with experimentation. Experimenters may "compare notes" and advise each other on how to interpret or solve problems, and this generates a more generally shared set of expectations for how to understand any given experiment. As Collins and Harrison (1975) amongst others note, this kind of social "rubbing up against one another" is an important element in the exchange of tacit and/or embodied knowledge and skill with respect to experimentation. While there is generally very little of this social rubbing during the formal talks at conferences, presenters can be "cornered" and informally questioned for details they may be unable to communicate in other forums. In addition, brief contact at a conference may lead to more in-depth communication or even collaboration amongst individuals and groups.

This latter point follows from a crucial second characteristic of meetings: they provide opportunities for participants to evaluate each other's character and hence credibility (Shapin 1994). Meetings as sites of social interaction often feature a fair amount of informal character assessment. In the case of cold fusion, with rhetorical displays of isomorphic replication creating the conditions for credibility assassination amongst participants (as the "difficulty" shifts from the experiment to the experimenter), face-to-face evaluations at meetings can have a profound impact on future indirect communication. Many scientists' judgments about the credibility of the cold fusion claims were affected by having met Fleischmann or Pons in person and by seeing the way they interacted with others. In some cases, an interaction with Fleischmann, for instance, might reduce a scientist's degree of interpretive charitability whereas in others it might increase. "I asked him a dozen questions and he wouldn't tell me anything," said one scientist to me over coffee at ICCF–4. On the other hand, according to another scientist, "It seemed to me that he was being completely sincere." On such occasions, doubts about the credibility of fellow researchers may be reduced or enhanced. This is partially a function of whether they can satisfy queries about their research, but it is also a function of the evaluation of moral character—their sincerity, honesty, openness, kindness, generosity, humility, and so on.[10]

Finally, the features of conferences that help ameliorate the conditions of the experimenter's regress and facilitate attributions of credibility (or incredibility) make conferences and meetings primary sites for identity making. Over time these meetings help produce common understandings of the phenomena in question, informal standards for evaluating the credibility of claims makers, and a sense of collective identification—a recognition of being a part of a common project with generally shared objectives and meanings. The dominant rhetoric at conferences may be celebratory, defiant or defensive, but interspersed with the exchange of technical information are statements in the introductions and conclusions of papers, speeches in plenary sessions, or the informal chat and gossiping between sessions that mark conference attendees to one another as members of the same group who share the same goals.

The difference between this normal social activity of conferences and the cold fusion case is that the production of researchers as a self-identifying group comes as a consequence of the end of a controversy in which all approaches to studying cold fusion (in any of its forms) are configured together as a form of pathological science. During the controversy in 1989, the interpretive flexibility surrounding the Utah results allowed for the phenomenon to be configured in different and legitimate ways motivated by scientists' background knowledge and interests in being able to produce the CF effects. With the closure of the controversy, cold fusion crystallized into a kind of nuclear phenomenon that did not (and could not) exist. The consequence of this was that all those who worked on cold fusion were perceived after 1990 as doing pathological science no matter what they thought the phenomenon might be, and this external identification became a source of collective identification and resistance on the part of CF researchers.

## *Building Solidarity through Resistance*

The first International Conference on Cold Fusion, held March 28–31, 1990, marked both an end and a beginning for CF research. The Salt Lake City conference was attended by two hundred researchers presenting some forty papers, most of which supported claims for the existence of cold fusion. The meeting was reported widely in the media, but the coverage was generally negative. CF researchers were referred to pejoratively as "true believers,"[11] and newspapers reported accusations made by the few remaining critics in attendance that conference organizers from the National Cold Fusion Institute (NCFI) had purposely excluded negative papers in order to make the case for cold fusion look more convincing. Taking place around the time of the publication of *Nature*'s final editorial and the negative isomorphic replication from Michael Salamon (March 29), the airing of a critical documentary produced by the BBC for the *Horizon* television series (March 26), and accusations of the misallocation of funds at NCFI (May), the cold fusion conference came to be publicly represented as more an act of desperation than an important turning point in the research effort.

For many cold fusion researchers, however, the opposite was the case, and what was interesting was the way in which reports of experimental data were framed by attempts to reconstruct and justify the field of research as a whole while reproving the unfair skepticism of critics and the media. Fritz Will, the director of the NCFI at the time of the conference, ended his own presentation with these comments:

> While key observations relating to cold fusion have been confirmed by many competent groups, it is also true that the phenomena cannot be reproduced on demand, and that an understanding of the underlying mechanisms is not at hand. The phenomena involve surface chemistry and the behavior of metal loaded with deuterium. Appreciating the complexities and well-known irreproducibilities involved in each of these cases individually, many scientists are not surprised that one year of research and development have not been sufficient to unravel the complexities of cold fusion. . . . We know that the reliable results obtained by a minority must not be regarded as wrong only because a majority of others has failed to confirm these results within one year.[12]

Will's comments helped to reconfigure cold fusion from the simple straightforward experimental entity many were expecting in the summer of 1989 to a more complex phenomenon requiring skill and patience to produce reliably. This facilitated the sense that the minority of scientists who had had success in producing results were more skilled and more patient than most others who were trying to replicate the CF effects, a notion that was the exact opposite of the judgment of the majority of participants in the controversy of 1989. Martin Fleischmann also emphasized this position in his presentation with a hint of defiance, but his comments ended on a note of concern:

> It's a nuclear process. We firmly believe this could only have a nuclear explanation. . . . We are convinced that these are nuclear processes, but we are not convinced that it is fusion. That's the substance of what we've showed so far. Our position is exactly the same as it was last spring. . . . I hope it's the end of the beginning and not the beginning of the end. I hope there is now a sufficiency of work which allows us to move forward to a comprehensive and rational design of new experiments. And I hope, in the end, that maybe something useful will come of it.[13]

Will's and Fleischmann's comments are certainly indicative of their sustained disagreement with other scientists, but in 1990 the nature of the disagreement had shifted. Both Will and Fleischmann had begun to recognize that their interests were being increasingly isolated and the prevailing scientific opinion had gone against cold fusion. The controversy had ended, and the research effort was in danger of falling apart. Instead of going through the normal practice in scientific controversies of making and defending claims, CF researchers faced the problem of establishing their legitimacy before their claims would even be heard. It is for this reason that the events of 1989 were recast as an early phase of cold fusion research, with the Salt Lake City conference marking a new beginning. The sense that cold fusion was in danger coupled with the desire to regain legitimacy produced the conference as a social space where the presentation of experimental data took the form of a call to arms and an attempt to bolster solidarity in the face of closure. Indeed it is in the context of these meetings that the stigma attached to cold fusion has become a potent resource for collective identification as well as the coordination of experimental practice.

It may seem curious, for instance, given the stigma that organizers would even refer to their meetings as being about cold fusion. Surely one strategy for regaining scientific respectability would be to present new work under a different label, especially since there was no agreement that "fusion" was what was really going on anyway. In the conclusion to his account of the controversy, Pinch considers this same situation:

> Today scientists still continue to work on the phenomenon, some positive results are reported, and conferences are held on such topics as "Anomalous Phenomena in the Palladium-Deuterium Lattice." Indeed, the very labeling of the phenomena in this way reflects a feature familiar from other controversies where, in order to try to get the controversial phenomenon accepted, proponents play down its implications for other scientists. Gone are the bold claims that the phenomenon is definitely fusion and gone is the prospect of a new source of commercial energy just around the corner. . . . It may eventually be established that there is something unusual going on in the palladium-deuterium lattice but that something is unlikely to be cold fusion as it appeared circa March 1989. (Collins and Pinch 1993, 77)

In this account Pinch draws on the epistemological contingency of closure to argue that the continuation of research and a resurrection of cold fusion may be possible in a different form under a different name. The implication of renaming the phenomenon is that researchers might be able to avoid being pathologized by dissociating themselves from the label "cold fusion." While he is partially correct about this, Pinch's analysis overlooks the dilemma inherent in the naming of the most important and widely attended meetings—the International Conferences on Cold Fusion.

The problem of naming is very much a feature of cold fusion research as undead science. On the one hand, Pinch's assessment is accurate. A number of researchers have purposely tried to bypass the stigma by adopting such labels as "chemically assisted nuclear reaction," "anomalous effects in deuterated materials," or "new hydrogen energy." In many instances the labels are more accurate descriptors for the phenomenon, but there is no doubt that they may also be safer labels. As Tadahiko Mizuno writes of his own experience, "When I submit a paper to a journal with the term 'cold fusion' in it, or I use that term in conversation with other scientists, I instantly sense their unease. I can see they are thinking 'something's fishy.' From that moment on they refuse to deal with me squarely" (Mizuno 1998, 110). Many others share Mizuno's experience, and this is not simply paranoia. Indeed in attempts to publish papers in mainstream journals or get approval for patent applications, "cold fusion" has become a kind of keyword signaling immediate rejection. While renaming does not assure publication, it does seem to help mollify negative reaction in some cases. Yet for all the attempts to rename the phenomenon, nothing has come into widespread usage to replace "cold fusion," and this is partly because the label of cold fusion still seems to accomplish some useful social work.

Part of what is occurring is a kind of collective identity formation that Castells (1996) terms "resistance identity" building. Castells defines resistance identity, following Calhoun (1994, 17), as a form of identity "generated by those actors that are in positions/conditions devalued and/or stigmatized by the logic of domination, thus building trenches of resistance and survival on the basis of principles different from, or opposed to, those permeating the institutions of society" (Castells 1996, 8). There is nothing new here. Calhoun and Castells are following in a long line of sociologists who have worked on issues of identity formation and resistance.[14] From this we can posit that the cold fusion conferences serve as the kind of "trenches of resistance and survival" Castells refers to, but with some important differences.

A crucial difference is that the identity building of cold fusion researchers does not fit Castells's ideal type because many researchers do not find themselves completely excluded, nor do they see themselves in wholly oppositional terms. As we have seen, a significant number of CF researchers are scientists working in respectable institutions and participating, for the most part, in regular scientific life. As resources continue to dwindle, cold fusion work increasingly moves to the back burner for these scientists, who must devote their time to more pressing and fruitful research. It is in fact quite rare to find a full-time cold fusion researcher, and thus CF work often involves the management of multiple identities as researchers move from their CF project to other projects and back again. (I give an example of this experience in chapter 6.) The consequence is that resistance identification amongst CF researchers is at best incomplete and somewhat ambiguous.

The ambiguity and fluidity of scientists's identities can be witnessed in their practices with respect to naming the phenomenon. When I asked one electrochemist whether he used the term cold fusion to describe his work, he told me, "Well . . . I do actually, in conversation. . . . I don't on a piece of paper [laughs], only because I don't know whether it is cold fusion. I gave a seminar last week called 'cold fusion research at [name of institution]' in fact it was called 'deuterated metals research at [name of institution]' but the code word is understood.[15]

This scientist's understanding of "deuterated metals research" as a kind of code points to the ways that some researchers manage and share more or less secret scientific lives. This duality is often expressed in interactions with administrators, senior colleagues, and granting agencies. One engineer outlined for me his strategy in writing papers so as to get them approved by his "bosses," who might be embarrassed or worried about his association with CF research: "First it's very detailed, and second it's not saying it's cold fusion it's saying it's something curious we have to look at it more or less—of course I call it cold fusion because that's the objective."[16] While CF researchers working in scientific institutions tend not to express their resistance identity in the workplace, they often do so in conversation with other CF researchers and at CF conferences. Indeed, one thing researchers can be reasonably sure of is that most cold fusion work will get a fair hearing at a cold fusion conference. The continued use of the cold fusion label partially attests to this, and one of the characteristics of undead science is this kind of double life for scientists: on the one hand closeting their work on cold fusion and occasionally expressing embarrassment or concern when the topic comes up with outsiders,[17] and on the other hand being open and expressive about CF work in contexts recognized as being safe.[18]

## *Revisiting Negative Replications*

Another important aspect of resistance identification in CF research is the revisiting and reanalysis of prominent negative replications from 1989, accompanied by the articulation of differences between the competence of early attempts at replication and post–1990 work. CF researchers have employed two different strategies for engaging with negative replications. One is to reconfigure old negative replications as null results (or incompetent experiments). Referring to the negative replications reported by scientists at Caltech, MIT and Harwell, one CF researcher writes the following in his literature review of the field: "In some cases, the conditions those studies used are now known to prevent the cold fusion effect. MIT researchers, for example, used an experimental apparatus that was open to the humid Massachusetts air and therefore subject to contamination by ordinary water, which has since been found to inhibit the cold fusion effect. Also, early experimenters used commercially available palladium without regard for its condition; it is now known that off-the-shelf palladium does not meet the special conditions required for cold fusion" (Storms 1994, 20).

At the same time, other CF researchers have reanalyzed the data presented in negative replications and argue that there were real anomalous effects that the replicators had neglected or dismissed as experimental error. This kind of reanalysis attempts to turn negative into positive replications. As one CF researcher told me,

> The most compelling positive experiments, the most convincing, in a perverse way, are these three from 1989: MIT, Caltech and Harwell. As you probably know, all three were claimed to be negative. Perhaps a better choice of words would be "proclaimed" or "advertised" as negative. Also, without a doubt, all three used terribly sloppy, third rate procedures, the wrong kind of equipment. All three declared their results before they finished the experiment. But, as it turned out, MIT was positive, but fraudulently misrepresented it as negative, Caltech was positive, but they made a dumb mistake in algebra and overlooked it. . . . when the three most powerful enemies of cold fusion run deliberately sloppy, half-assed experiments and still get positive results, that— by golly—is the best possible proof that the effect is real![19]

In this comment, the rhetorical isomorphism of the negative replications is turned back on itself, facilitating the representation of the replicators as "dumb" and "sloppy." Martin Fleischmann, who has spent much of the past few year reanalyzing old data, takes a similar view. In a recent interview for the cold fusion magazine *Infinite Energy*, he states, "If you take the Harwell data sets, you cannot say that this experiment worked perfectly and that there is no excess heat. You could only say either that the experiment worked perfectly and there is excess heat or the experiment didn't. And on those two bases you have to do another set of experiments. As regards MIT, all one can do is shake one's head in disbelief really. I mean, again, if you fiddle about with baselines then you have to consign those experiments to the dustbin and start again. The one in Caltech, was clearly very strange because there was a redefinition of the heat transfer coefficient" (Tinsley 1997, 14). According to

Fleischmann, the Harwell replication produced a positive result that the replicators failed to see because they did not perform the correct analysis.

In this post-closure period, however, the scientists at Caltech, MIT, and Harwell do not seem to care or see a need to disagree with this new interpretation of their experiments. The attempts to reanalyze old negative data have been met largely with silence. CF researchers make sense, and take advantage, of this silence by arguing that the critics from 1989 now irrationally refuse to even consider the new data: "I've gone over, I've talked to [an MIT physicist] about my ideas about the field. [The MIT physicist] doesn't have to see any of the papers, doesn't have to see any of the experimental results; he doesn't have to see anything. He stops after the first two sentences—gosh you think you can get something nuclear with electrochemistry—I'm sorry but I know the answer to that—there's no effect."[20]

One consequence of this silence and continued rejection by critics is that along with a reconfiguration of negative replications comes a reconfiguration of skeptics' positions as a kind of "pathological skepticism" in which disbelief in cold fusion is reduced to the irrationality and irresponsibility of a small group of individuals."[21] In a letter to his colleagues at a research laboratory in India, one chemist writes, "I am amazed to see the intense anti–cold fusion propaganda launched by a handful of influential people at an international level. Most of the highly critical articles on cold fusion can be traced to a few writers. . . . most of these writers do not refer very conveniently to the large number of confirmatory results which have poured in from different parts of the world in recent months."[22] Here, the network of opposition to cold fusion is reduced to the opinions of an influential handful of writers who ignore new experimental evidence. The chemist's letter, aimed at convincing colleagues to reconsider the case for cold fusion, rhetorically reweighs the outcome of the controversy, finding that the number of confirmatory results is "large" compared to the "handful" of critical articles. While this letter was not particularly successful in terms of resurrecting cold fusion work at the author's institution, that is not all that is going on, as the author also uses the letter to define himself as part of "a small band (estimated to be about 600 scientists at present) of dedicated researchers . . . [who] have continued to study the subject passionately and have indeed obtained, at least in my judgment, very convincing evidence of the authenticity of the phenomenon and its nuclear origins."[23] In this sense the letter becomes important for the author in affirming his own identification with the dedication and perseverance of a community of practice beyond his own institution.

The pathologization of skeptics, like the pathologization of CF researchers, provides an explanatory resource to help researchers make sense of their situation. Like the thesis of pathological science, the continuum of interpretation includes a notion of fraud. At the extreme limit of pathological skepticism is conspiracy. This is the claim, often discussed informally by CF researchers, that powerful elites in the U.S. Department of Energy, the patent office, major research laboratories, and the oil industry are responsible for the continued repression of CF research. My own interest lies not in authenticating accusations of conspiracy but rather in examining the social effects of "conspiracy talk" specifically and talk about pathological skepticism more generally. The same kind of boundary work that facilitates the stability of conventional nuclear physics by marginalizing cold fusion research also facilitates the stability of cold fusion by marginalizing skepticism. The process helps to stabilize an identity for both the CF researchers and the object they study.

## *The Ambiguity of Fleischmann and Pons*

One of the consequences of the pathologization of cold fusion research has been the almost total dismissal and neglect of CF work performed and reported by researchers other than Fleischmann and Pons. In the process of pathologization Fleischmann and Pons's incompetence became the cause of cold fusion, and cold fusion was reduced to the claims of Fleischmann and Pons. It is for this reason that undead science is sometimes difficult to locate. In the writings of boundary workers and in media reports, few names other than Fleischmann and Pons are ever associated with cold fusion, and it is easy to see how resistance might come to an end once Fleischmann and Pons are gone. Yet such perceptions are a consequence of pathologization; scapegoating may help normalize scientific knowledge and secure the boundaries of social worlds, but it is not an especially useful tool for doing the sociology or history of science. Part of what I have been trying to argue in this book is that CF work continues in a community of scientific practice that extends well beyond Fleischmann and Pons or even their charismatic influence. At the same time, however, it is important to understand that this world takes its shape from researchers' experience of cold fusion's identification with Fleischmann and Pons.

At the end of 1990, things were not going well for Fleischmann and Pons. They had published a long, detailed paper addressing the calorimetry of their CF cells (Fleischmann et al. 1990), but this did not seem to have any effect on the prevailing opinion. A collaboration between the University of Utah and the Los Alamos National Laboratory fell through, a partnership with General Electric failed to materialize, and the University of Utah was demanding access to Pons's laboratory records as part of an investigation into the operation of the NCFI. In January 1991, Pons resigned from teaching at the University of Utah and went on extended leave. Fleischmann was suffering from colon cancer and after an operation returned to Southampton to convalesce. All of these events appeared in occasional media reports and the writings of boundary workers, which further supported the thesis of pathological science, reducing cold fusion to the imputed improprieties of Fleischmann and Pons and others at the University of Utah.

In 1991, however, Fleischmann and Pons had a change of fortune. After almost a year of not being heard from, they reappeared in Sophia-Antipolis, a scientific research park near Nice, France. They were working in a laboratory operated by IMRA Europe (an R&D subsidiary of the Toyota Motor Company), with funding from a Japanese speculative research consortium called Technova.[24] While there were lingering problems with the University of Utah, it seemed that at the IMRA lab, Fleischmann and Pons were working under better conditions and with more support than ever before. Under the terms of a five-year renewable contract, Fleischmann and Pons were to assist Japanese researchers in isolating the critical parameters for developing the commercial viability of cold fusion. Despite the outcome of the controversy in North America, Europe and even Japan, senior managers at Technova, particularly Minuro Toyoda, had decided cold fusion research was worth supporting.

It was support from Technova and IMRA that allowed Fleischmann and Pons to conduct their heat-after-death experiments, and it was rumored that they were also trying to develop a CF demonstration device, something like a water heater that could provide clear evidence of over-unity excess energy.[25] Fleischmann and Pons's change of fortune meant that they continued to run the most amply funded research project of any group working on CF, but their deal with Technova also brought a new level of secrecy to their work. Consequently it became increasingly clear to many researchers that they would not be able to count on Fleischmann and Pons to resolve the problems of reproducibility and legitimacy that plagued the field. Alongside the IMRA project, CF research continued along multiple trajectories in numerous other groups around the world, some of which approached the topic from wildly different perspectives.

In the end, the work at the IMRA lab did not live up to its promise. In 1995 Fleischmann retired from the Technova project and moved back to England. Pons remained until 1998, when the laboratory was finally closed (the Technova contract was not renewed), and although nearly $40 million had been spent, no viable commercial product or demonstration device had been developed. Fleischmann maintains that their research progress had been hindered by the commercial demands of Technova management but that substantial innovations had indeed been made at a more basic level (such as the work on heat-after-death). While Fleischmann continues to attend meetings and give talks, Pons has remained silent, and as of 2001 their experimental research is at a standstill. The two men published over twenty papers on CF in the past twelve years, and Fleischmann especially is still widely respected as one of the founding fathers of the field. At the same time, it has been important for the survival of CF that research has developed independently of Fleischmann and Pons's influence.

Establishing closure in 1989 involved the successful association of cold fusion with the experimental incompetence of Fleischmann and Pons. As long as cold fusion was being produced only in Fleischmann and Pons's laboratory, their fate and the fate of phenomenon could be linked, making cold fusion that much easier to kill. In the post-closure world, cold fusion survives through the production of CF effects in many locations by many different researchers in a manner that resists the reduction of CF to the "failure" of Fleischmann and Pons (since every criticism of Fleischmann and Pons is more than balanced by the "successful" replications by numerous other researchers). But there is a strange irony in this form of post-closure survival. Fleischmann and Pons's method for producing CF effects does not work, in the sense that other researchers have difficulty using their method to produce the effects consistently for themselves. These problems with reproducibility combined with the greater secrecy as a result of the Technova contract and the public stigma associated with Fleischmann and Pons to produce a variety of alternative experimental methods and even new effects that have been difficult to reconcile. The irony is that what holds these alternatives together and allows researchers to recognize data from different experiments as being indicative of the same phenomenon is their common association with Fleischmann and Pons and the label of cold fusion. Thus Fleischmann and Pons and the label of cold fusion make it more difficult to do post-closure research but also provide the conditions for collective identification that make coordinated action and belief amongst CF researchers possible.

## *Making Cold Fusion Work*

It is beyond the scope of this book to provide a comprehensive review of all the technical developments, issues, and debates that have occurred in the ongoing research on cold fusion over the last twelve years. To date over 1,200 articles related to cold fusion have been published in a variety of scientific journals, and much more information has been reported in conference papers and proceedings, patent documents, technical letters, newsletter and magazine articles, and reports on the Internet. Despite the difficulties researchers have in publishing their work, this represents a substantial output for a twelve-year period. As a consequence of this volume of work, it is difficult to describe and summarize all the trajectories of research on cold fusion, a situation made even more difficult because of the loose associations and dispersal of CF researchers around the world. At any one time there are several dozen groups around the world engaging in varying degrees of research activity, and as this book goes to press new data from different experiments are reported every few months.[26]

Some of these activities are neglected or dismissed by the cold fusion community, others spark flurries of research activity, and a few prompt wholesale reevaluations of the field, but in fact very few CF researchers manage to keep track of all developments, and there is a recognized need for literature review articles that can synthesize data from various reports with a view to developing more common ground in the numerous research efforts that manage to survive. Edmund Storms, a retired radiochemist who worked at the Los Alamos National Laboratory, has authored two of the most cited literature reviews (Storms 1991, 1996a) and currently there is one book (Hoffman 1995) providing detailed technical criticism of cold fusion experiments up to 1992 or so and another recently published book (Baudette 2000) providing the most up-to-date overview. The account that follows is drawn from these works, in addition to conversations with a number of CF researchers. My intention is to describe those elements of experimental practice and understanding that have come to be shared by most researchers still doing work today.

Over the last twelve years cold fusion research has been defined by what might be called a core methodology more or less following Fleischmann and Pons's 1989 approach. This line of investigation has depended on building various configurations of electrolytic cells using different shapes and sizes of mainly palladium but also nickel and titanium cathodes. This core "wet electrolysis" approach is supplemented by other techniques aimed at bringing deuterium or hydrogen ions into interaction in the presence of a metal catalyst. Experiments using these other methods have also produced a variety of CF effects, including excess heat and nuclear ash (helium–4, neutrons, and tritium). A few of the other methods include:

a) Gas or plasma discharge. This is a set of techniques that use deuterium gas or plasma instead of liquid $D_2O$. Palladium cathodes (or sometimes titanium or nickel) are placed under high pressure (and sometimes temperature) in the presence of deuterium gas or plasma, which may be subjected to a low-voltage current or high-voltage sparking. Another variation of this method uses plasma discharge in $D_2O$ or $H_2O$.

b) Ion bombardment. This technique uses a particle gun to shoot a high-intensity beam of deuterium ions at a palladium or nickel target.
c) Cavitation. This technique makes use of sonoluminescence, in which tiny bubbles of liquid are made to cavitate (grow and collapse) using high-intensity acoustic energy (ultrasound). In some experiments bubbles of $D_2O$ are directed at a palladium (or other metal) foil target, producing sonofusion.

For most of these alternative techniques, the principle of the experiment remains the same. The main operating assumption is that in order to produce CF effects one must find a method for facilitating or forcing the interaction of deuterons with a metal catalyst. Gas-phase experiments are particularly useful for this in that they avoid problems associated with liquid electrolysis. In electrolysis systems, deuterium loading is a notoriously uneven process: sections of the metal may absorb a higher density of deuterons, causing cracks to form, which may result in the deloading of deuterium. Under the controlled temperature and pressure of gas environments, however, loading tends to be more uniform, and cracking appears to be less of an issue. The drawback of this method is that the apparatus requires more expensive equipment, and it is more labor intensive and costly to set the system up and run it. So while data from some gas-phase experiments appear to be more robust and more consistently reproducible, they are also harder to replicate by others and more importantly are difficult to integrate with the results of the wet-electrolysis techniques, at least with respect to calculating anomalous heat (since the heat balance calculations differ significantly between the two kinds of experimental systems).

Some researchers argue that the CF community should abandon Fleischmann-and-Pons-style electrolysis in favor of this other approach. Yet as a number of researchers told me, this has been difficult because so much effort has already gone into developing Fleischmann and Pons's basic method, and researchers see the vindication of cold fusion as depending on answering criticisms based on this original method. This situation constitutes another tension in post-closure CF research. Do scientists continue working with and improving on the more familiar Fleischmann and Pons protocol, or do they concentrate on potentially more promising techniques that fewer participants will be able or willing to work with?

## *(Im)proving Excess Heat*

Since 1989 many researchers have argued that the strongest evidence for cold fusion comes from the measurement of excess heat or power in experiments. The problem with this, as we have seen, has been that as far as physicist critics are concerned, the only definitive indicator of nuclear reactions taking place is the presence of some reaction byproduct like neutrons, gamma rays, tritium, or helium. The response from Fleischmann and Pons and others has typically been that while it is important to search for these byproducts, it is also important to understand that the magnitude of excess energy being detected is far greater than can be accounted for by any normal chemical or mechanical process. The CF effects are thus seen to be nuclear by default (although there has never been any agreement that the effects are specifically due to fusion). While a few scientists have considered the possibility of unknown or esoteric chemical and mechanical processes, most CF researchers maintain that the phenomenon occurs at a nuclear-physical rather than chemical scale. Thus experiments directed at producing excess heat have focused on more precise calorimetry and attempts to increase the magnitude of the effect (using different materials and cell configurations).

One of the earliest criticisms of Fleischmann and Pons's experiments was that in using a cell design open to the surrounding air, they could not account for energy that might be produced in the recombination of oxygen and deuterium gas during electrolysis. Fleischmann and Pons argued that even if all the ions of oxygen and deuterium recombined in their experiments (which they didn't), the total energy of recombination would still not add up to the excess energy they were measuring. From their point of view, to obsess about recombination was merely nitpicking. Fleischmann and Pons ran a control cell using platinum instead of palladium cathodes, pointing out that even though recombination would be a factor in both the platinum and the palladium cells, only the palladium cells produced excess heat. Their arguments made some sense but in not simply capitulating to the request for closed-cell calorimetry, they missed an opportunity to ease the tension between CF researchers and skeptics.

In part to address criticisms about recombination and in part to improve the precision of their calorimetry, many CF researchers started using closed-cell calorimeters equipped with recombiners. In addition, emphasis was placed on the use of flow-type calorimeters (as opposed to Fleischmann and Pons's "static" calorimeter), in which the electrolyte moved through the cell one way and out another. The advantage of this was that temperature measurements could be made outside the electrolysis chamber, thus removing any error due to thermal gradients in the cell or other random effects. Finally, insulated calorimeters minimized the exchange of heat between the inside and the outside of the cell. All of these features increased the precision with which scientists could measure heat production in the cell, and the majority of the most well regarded positive replications since 1990 have used this more sophisticated instrumentation. The increase in precision has resulted in more reliable anomalous-heat measurements, with excess-power results well beyond the threshold limits of the measurement systems used.

Perhaps the most highly regarded research program over the last few years has been led by Michael McKubre at the Stanford Research Institute. McKubre trained under Fleischmann at the University of Southampton, and while he started experiments along with others in the spring of 1989, he kept relatively quiet about his research until 1990.[27] McKubre's research, like the research at Texas A&M, was initially funded by the Electric Power Research Institute and later by IMRA, the national hydrogen energy project of the Ministry of Japanese Trade and Industry (MITI), and most recently the U.S. Defense Advanced Research Projects Agency (DARPA). McKubre's research is generally viewed as being comprehensive and sophisticated. He often appears at CF meetings, where he seems to lend an air of conventional respectability and humility to the proceedings with his sober presentations. Most importantly, CF researchers generally consider McKubre's calorimetry to be beyond reproach, making his anomalous-heat results especially convincing. For this reason, his work is often cited in CF researchers' attempts to rebuild legitimacy for the field. McKubre is also convinced of the quality of his group's calorimetry, but he has received very little attention from mainstream scientists: "One of the attempts we made to validate our method of calorimetry was, in fact, to take the design and the results, including excess power results to an annual calorimetry conference and expose it to the experts. . . . They didn't tell us that anything was wrong with it by and large. Did we receive acclaim? No. But a rather grudging acceptance of the mode of calorimetry that we had developed. As yet nobody has ever told me of an issue or problem with the mode of calorimetry that we developed. Which is one reason we stick to it."[28]

It should be noted that one reason for the lack of attention to McKubre's work is that before 1994 detailed reports of it were difficult to come by (as the research occurred under the auspices of a private research institute and a commercial enterprise), and by the time McKubre's final EPRI report was released, there were few scientists who were not CF researchers who were willing to take the time to look closely at further cold fusion experiments.[29] Despite the lack of critical attention, CF researchers tend to agree amongst themselves that McKubre's technique addresses all the major criticisms of excess-heat results from 1989.

## (Im)proving Nuclear Ash

Far less attention has been directed to the measurement of nuclear ash than to the measurement of excess heat. This may be partly because many post-closure CF researchers have more experience with calorimetry, but it is also because calorimetry involves more accessible and less expensive instrumentation than neutron or gamma-radiation detectors. Two other factors also influence a lower interest in nuclear ash. Many researchers believe that the anomalous energy effects are more stable and more reproducible than the nuclear ash effects. Reported measurements of neutrons, gamma radiation, helium, and tritium varied greatly in 1989, and without a viable theory for the phenomenon, it made more sense to concentrate on excess heat as the most reliable signature of the effect. This was the exact opposite of the position of the DOE ERAB panel, for instance, which viewed nuclear ash as the most important signature. The other influential factor has to do with the fact that excess energy and not nuclear ash is the more important effect in terms of developing a commercializable technology. From the point of view of commercialization, it really didn't matter if the phenomenon was nuclear at all as long as it held the promise of being a clean, cheap, and efficient energy source.

Nevertheless, important innovations have been made in the detection of nuclear signatures in CF experiments. Carrying the dubious honor of being one of the cofounders of the field, Steven Jones continued his work on low-level neutron emissions from CF-like cells until 1995. As with McKubre's sensitive calorimetry, Jones and his colleagues spent a great deal of time developing an innovative neutron detection system, which they located deep underground to minimize background radiation. In Jones's case, the more sensitive his detector became, the less convinced he was that he had measured anomalous neutron emissions in his cells. In 1994 he officially retracted his neutron claims, and after engaging in critiques of other CF researchers he found himself somewhat marginalized within the CF community as a whole. Jones has argued that without positive results using sophisticated and sensitive instrumentation, belief in the phenomenon cannot be justified. He has even extended an open invitation to other researchers to test their active cells in his neutron detector, but while some discussions of collaboration have taken place, no one has taken Jones up on his offer. Despite Jones's position, there are other researchers who are confident that their neutron detectors are good enough, and throughout the 1990s low levels of anomalous neutrons and neutron bursts have been reported by groups in the United States, Japan, Italy, and India. The problem has been that neutron effects are not reliable, appearing without any regularity in cells that are actively producing heat.[30]

More fruitful research, on the other hand, is being carried out on detection of tritium and helium in CF cells. As I explained in chapter 4 while discussing the case of John Bockris and Nigel Packham at Texas A&M University, anomalous-tritium results in particular have been dogged by criticisms that cells had been contaminated, or as Gary Taubes claimed, "spiked." Anomalous tritium has been reported in several experiments, but some of the most convincing results come from Thomas Claytor, a physicist working at the Los Alamos National Laboratory.[31] Claytor's results are considered significant because of his expertise and access to the sophisticated equipment at Los Alamos, but they have also proved to be inconsistently reproducible, and he has not measured excess heat in any of his tritium-producing experiments.

Similar problems exist with helium–4 detection. Helium–4 exists in relative abundance in the air and can easily contaminate cells, so that great care must be taken in the handling of samples, which are usually sent out to private laboratories for analysis. Melvin Miles at the China Lake Naval Weapons Research Center in California performed some of the most highly regarded helium experiments during the 1990s. Miles collected the gases that bubble off during electrolysis experiments, generating excess heat, and sent them to other laboratories for blind analysis using mass spectroscopy.[32] In one set of experiments, six out of six cells producing excess heat also produced anomalous helium–4, while eight out of eight cells that did not produce heat also did not produce helium–4. In addition to this, Miles reported that the amount of helium could be correlated to the excess heat measurements such that the reaction $D + D \rightarrow 4He$ + heat might be the prevailing reaction.

Like McKubre, Miles has received little attention outside the cold fusion community. In 1995, after a brief exchange of criticism by e-mail that appeared in the USENET newsgroup sci.physics.fusion, Steven Jones and Lee Hansen published a critique of Miles's experiments in the *Journal of Physical Chemistry* (Jones and Hansen 1995), which was followed three years later by a rebuttal from Miles and a final reply from Jones (Miles 1998; Jones, Hansen, and Shelton 1998). Jones and Hansen criticized Miles's calorimetry as well as his controls and accused him of not taking proper account of recombination as a source of excess heat. Miles defended his claims, and while there appeared to be some confusion in communication between the scientists, once again researchers' competency was in question and tempers flared (a situation not helped by the editor's delay in publishing Miles's rebuttal). Most researchers supported Miles as the distance between Jones and the rest of the CF community grew larger. As has been the case with other CF effects, Miles has not been the only researcher who has consistently measured helium. Experiments by Daniele Gozzi at the University of Rome and later by Y. Arata and Y. C. Zhang have confirmed helium–4 at levels close to those described by Miles.[33]

At least with respect to published reports and conference presentations, there has been a significant increase in the care and complexity with which measurements of effects in many CF experiments have been performed. This has led most experimenters who had success in 1989 to become more rather than less convinced after 1990 that the phenomenon is real, despite ongoing difficulties in reproducing their own anomalous results consistently. In short, although the effects cannot be produced reliably, when they do appear they tend to be more robust. As CF supporters often note, this is at odds with the symptoms of pathological science. The error bars on plots of experimental data are smaller, and systematic sources of error are more easily accounted for and eliminated. Experimenters are careful to use controls (either a plain $H_2O$ or a $D_2O$ electrolyte with a platinum cathode, or a "dead" palladium cathode), and some have the resources to use redundant instrumentation. All these innovations give researchers the sense that CF experiments are "better" today than they were in 1989, and there is general agreement about what count as better or worse experiments, although in many cases CF experimentation is still a matter of making do with resources that are available.

## Critical Loading

In addition to the development of more sophisticated measurement systems, perhaps the most pressing concern has been the effort to figure out how to produce CF effects reliably and consistently. Along with calorimetry, McKubre's SRI group is responsible for bringing another issue into bold relief amongst CF researchers. McKubre, like other electrochemists who worked on CF, was already used to working with metal hydrides and electrolytic systems in 1989. Since he had never noticed any anomalous effects in these systems before, his early experiments were directed at what he suspected were the particular conditions that might produce the reported CF effects. Like a number of other researchers early on, McKubre hypothesized that only once the palladium cathode was saturated with enough deuterium would any effects occur and that the failure to observe CF effects was due to insufficient deuterium loading in the cathodes of would-be replications.

McKubre's group was one of the first to correlate the loading ratio (the ratio of deuterium ions to palladium atoms in the cathode) with the production of anomalous energy. By 1994 he was reporting that cathodes with ratios that were not close to 1:1 D:Pd never produced anomalous energy, while cathodes with high ratios produced excess heat more often than not.

The discussion of loading ratios had two important effects on the development of CF research. First, it provided a readily available explanation for the failure of so many attempts in 1989 to detect excess heat. It turned out that deuterium loading in palladium is not a simple process. If the cell is not electrolyzed long enough, there may not be enough deuterium in the cathode to generate high loading. If the current density in the cell is too low, deuterium may leak out of the cathode, making high loading difficult to achieve. In addition, loading deuterium into palladium tends to stress the metal, and this may result in small cracks through which deuterium can escape. The consequence of this was that different groups found they could get 100% reproducibility with cathodes made from one batch of palladium metal but not with cathodes made from another batch (even if these were from the same company). Since the majority of negative replications did not provide a measurement of the D:Pd ratio, CF researchers were able to discount these experiments while finding a plausible explanation for inconsistent reproducibility in their own experiments.

The other major consequence of McKubre's highlighting of the loading issue was to crystallize efforts to focus on characterizations of the kind of metal-hydride system under investigation in CF work. Researchers seem to be in general agreement with respect to a number of critical parameters for getting their CF cells to produce results (at least in the cases where $D_2O$ electrolysis is used). These critical parameters are well summarized by Edmund Storms, who proposed a basic procedure for producing CF effects aimed at those having trouble:

> This procedure requires knowledge of the cited literature and some experience constructing and using electrolytic cells of this type.
>
> 1. Polish the palladium surface to a mirror finish and wash with acetone. Do not touch with fingers or tissue after this step.
>    Purpose: To remove impurities, to produce a uniform surface, and to create a dislocation layer.
> 2. Heat in air at 600–700 C until the surface oxidizes to a uniform blue color. This operation is best done in a small furnace. If the color is not uniform, repeat step 1.
>    Purpose: To remove impurities, to demonstrate uniform chemical activity of the surface, and to produce a layer of pure palladium on the surface.
> 3. Weigh and measure dimensions.
> 4. Electrolyze as the cathode in 0.3–0.5 LiOD electrolyte at 20 mA/

cm$^2$ until weight increase indicates a composition greater than $dD_{0.75}$.

5. Remeasure dimensions and calculate excess volume. Discard if excess volume is greater than 2%. "Good" material may be stored in a sealed container below 0 C indefinitely.
   Purpose: To eliminate palladium having cracks and voids after hydriding.
6. If excess volume is less than 2%, place the palladium in a prepurified cell and place this cell in a calorimeter. Electrolyze the cell at 200–500 mA/cm$^2$. If excess energy is not produced within 300 hours, start initiation procedures.
7. The following initiation procedures can be applied in sequence or at the same time:
   a) Add 10ppm of aluminum metal to the electrolyte. Electrolyze for an additional 100 hrs.
      Purpose: To create a barrier to reduce deuterium loss.
   b) Switch electrolytic current between 20 mA/cm$^2$ and 500 mA/cm$^2$. The switch rate can be between 1/sec and 1/min. Continue switching for an additional 100 hrs. Test for excess energy while the current is fixed at 500 mA/cm$^2$.
      Purpose: To create a variable flux of diffusing deuterium.
   c) Increase the temperature to at least 50 C and continue to electrolyze at 200–500 mA/cm$^2$. This step is best done after heat production starts.
   d) Inductively couple an RF frequency of 81.9 MHz to the cell through a surrounding coil. Power levels below 30mW have been found to work well but values up to 1 mW can be used. If excess energy is to be produced, it will appear immediately.

If none of these techniques work, repeat the process using a new piece of pretested palladium. (Storms 1996b)

The idea of this kind of recipe is to stabilize practices involved in producing CF effects in wet electrolysis Pd-D systems. The main difficulty, according to Storms and many others, is actually finding "good" or active palladium samples, and it is not clear what characteristics of a specific sample make it work. On this issue, chemical manufacturers have had an important role to play in the life and death of cold fusion. The company that supplied Fleischmann and Pons with their active samples of palladium, Johnson and Matthey, no longer manufacture palladium in the same way (the purification process is different), and there are no more batches of the original palladium left. Other attempts to produce active samples, by Tanaka Metals in Japan, for instance, have had reported successes, but these have not been reproducible.

Yet despite these perceived difficulties, there is still some shared understanding of how to recognize an active cathode once one has it. Storms suggests that the detection of cold fusion shares properties with the detection of superconductivity as a special condition of matter:

> The claimed effects appear to occur in special environments. For lack of a better designation, these environments will be described as a special condition of matter (SCM), perhaps similar in some respects to the superconducting state. . . . One such proposed SCM can be formed after beta-PdD has achieved a sufficiently high deuterium content. Once this essential requirement has been achieved, several other conditions are required to form the SCM. These include the presence of certain impurities in the palladium and the application of various forms of energy. Once a stable and sufficiently high D/Pd ratio has been produced, excess energy production starts after many additional hours of electrolysis. (Storms 1996b)

Like the Meissner effect in superconductivity, high deuterium content in Pd (amongst other things) becomes a necessary test for cold fusion. To the extent that other researchers agree with Storms's position, a collective criterion for recognizing cold fusion (thus differentiating the genuine anomaly from experimental artifact) begins to take shape. Experiments that produce anomalous effects in the absence of high loading are considered to be of lesser quality than experiments producing anomalous effects under conditions of high loading.

## *Divergences in the Shape of Cold Fusion*

But the process of shaping cold fusion is far more complicated than I have indicated thus far. Difficulties in comparing and integrating the data using different methods for producing CF effects are compounded by the fact that the effects being reported tend to vary across similar kinds of experiments. Generally speaking, differences in the measured quantity of excess heat or above-background nuclear ash are not of particular concern to CF researchers, since any experiment that produces any amount of unaccounted for heat or nuclear ash may be indicative of cold fusion. As a result, better experiments tend to be simply those that produce the most robust data (with measurements that indicate anomalous effects well above background), while the best experiments produce a combination of anomalous-heat and nuclear-ash data alongside data from control experiments. Major difficulties in interpretation arise, however, for experiments producing anomalous results that contradict one another. In this situation (assuming a researcher is aware of the contradiction) decisions are made as to the relevance of the conflicting data and the quality and competence of the experiments.

A good example of this is the case of light-water cold fusion (LWCF). In 1991 and 1992, a few research groups started reporting small amounts of excess heat and tritium in a variation of what had become the standard CF cell.[34] The variation used nickel in place of palladium and light water in place of heavy water.[35] While the levels of heat and tritium were lower than Fleischmann and Pons had found, this new development seemed exciting for a number of reasons. First, cold fusion in light water would be an even more revolutionary discovery than in heavy water, both in economic and scientific terms, and second, light-water experiments were far easier to perform, making the field more accessible to a wider variety of researchers. In the context of post-closure research, the possibility of light-water CF raised the stakes of the game as well as facilitating the participation of more researchers. Skeptics, on the other hand, viewed the light-water results as a disturbing ad hoc development, arguing that the inability to reproduce cold fusion had led researchers to posit ever more outrageous claims, especially given the debate about light-water controls in 1989 (see chapter 3).[36]

CF researchers working with standard D-Pd cells found themselves in a quandary. Fusion in D-Pd systems was already difficult for mainstream scientists to accept; fusion in hydrogen-nickel (H-Ni) systems would make the battle to regain legitimacy even more difficult. A more significant problem perhaps was that success with H-Ni systems would create difficulties for understanding what was going on in D-Pd systems. For one thing, the light-water results, if accepted, raised difficulties with using light water as a control for heavy-water experiments. More significantly, however, the light-water results also seemed to indicate that critical deuterium content might be less of an issue. Since nickel does not absorb hydrogen to the extent that palladium does, then perhaps high loading ratios are not a necessary condition for producing the effect.

One response of CF researchers to the H-Ni results was to question the competence of the experiments. When I asked one electrochemist in 1995 what he thought of the light-water–nickel research, he said, "I don't take that as being anywhere near proven. The calorimetry in general is of much inferior quality to the calorimetry performed in the heavy water studies. The likelihood is much less. . . . my gut feeling is that a deuteron is more likely to enter a nuclear process than a proton. . . . I don't take the light water data into my information set. . . . I don't think I need to explain it."[37]

But other CF researchers view the light-water systems as an important innovation, arguing that they produce more robust results: "Noteworthy features of the Ni-$H_2O$ cells, as described by those who have experimented with them, are: (a) they have very short initiation times, i.e. the 'excess power,' if present, appears within the first day of electrolysis and (b) the success rate of observing 'excess power' is high compared to Pd-$D_2O$ systems. On the whole, the system appears to be more robust and easily amenable to experimental investigation" (Srinivasan 1994). Still others view the light-water results as indicative of a more radical nuclear process, since as unlikely as deuterium-based nuclear reactions in palladium might be, a hydrogen-based reaction is even more unlikely. Perhaps the process is not even nuclear but rather something more exotic.

Such a possibility is suggested by Randell Mills, a medical doctor turned energy researcher. Mills has reported some of the most robust results from light water experiments, and his company, Blacklight Power, is marketing his "HydroCatalysis" process. Mills has proposed a chemically based theory that the excess energy in CF experiments is due to a catalytic reaction in which the electrons of hydrogen atoms move to an energy state lower than the "ground state" as defined by normal quantum-mechanical models of the atom. In the process of electrons moving to this lower energy state, excess energy is released and a new form of hydrogen is created, which Mills refers to as a hydrino. While Mills's hydrino theory does not require a nuclear reaction to explain the excess heat, it does require a wholesale overhaul of quantum mechanics, a prospect that many CF researchers are not enthusiastic about. More important, however, is that Mills's theory does not seem to account for CF experiments that produce anomalous nuclear ash. Although Mills himself avoids associating his work with cold fusion, his theory is championed by a small group of CF researchers whose work remains somewhat sidelined in favor of heavy-water CF research.

While light-water experiments are generally seen to be less attractive to the core group of CF researchers, in 1992 reports of other sorts of CF effects were starting to generate some excitement. Researchers doing spectroscopic analysis of the cathodes in their cold fusion cells were detecting evidence of new isotopes and elements in excess-heat-generating cells. These researchers found that the distribution of isotopes of various heavy elements differed before and after experiments producing excess heat. The implication was that one of the possible effects of cold fusion was nuclear transmutation, a process whereby one isotope of an element is turned into another or into a completely different element. Tadahiko Mizunohas has reported some of the most compelling evidence for transmutation. Mizuno used palladium cathodes in Fleischmann-and-Pons-type cells as well as his own gas-phase variation using strontium-cerium proton conductors instead of palladium for the cathode material. Using a variety of different kinds of elemental analysis, Mizuno found evidence for the production of anomalous distributions of elements like platinum, chromium, iron, and copper, which he decided could not be explained by contamination. For one thing, the measured distributions of isotopes of different elements did not conform to the distributions of the naturally occurring forms of those elements. The measurements for copper, for instance, seemed entirely strange. In naturally occurring copper (Cu), the ratio of the isotopes Cu–63 to Cu–65 is about two to one, but in Mizuno's experiment there was no Cu–65 present at all. This implied that the copper present in the cathode sample was not a result of contamination (at least from naturally occurring copper). In at least five different experiments generating excess heat, Mizuno found evidence of isotopic shifts. These results were corroborated by Richard Oriani, an electrochemist working at the University of Minnesota with whom Mizuno had been exchanging data.

Mizuno's data joined similar reports from researchers in Russia, Japan, and the United States who were finding all sorts of strange isotopes of elements, and these reports have effectively expanded the field of research under the label of "low-energy nuclear reactions" (LENR).[38] For LENR researchers, energy release from fusion is viewed as only one possible kind of nuclear reaction that might be occurring in metal-hydride systems. Isotopic shifts serve as evidence for the occurrence of a nuclear reaction, but such shifts are less likely to be the result of a fusion process. Transmutation research occupies a position similar to that of light-water work within the larger world of cold fusion research. Claims for transmutation or isotope-shifting effects are considered even more heretical by critics and skeptics, who associate these reports with alchemy. As with the age-old alchemical dream, some researchers suggested that cold fusion might be used to produce precious metals like gold, or in a more appropriate twenty-first century dream, the suggestion is that LENR effects could be used to eliminate radioactive waste by transmuting radioactive isotopes to nonradioactive ones. As one might expect, the skepticism of those outside the CF research community has had palpable effects on the overall cold fusion research effort.

In December 1993, for instance, a petition was circulated amongst faculty at Texas A&M University to have John Bockris stripped of his "distinguished professor" status because of his work on cold fusion, especially his association with work on transmutation. The petition was strongly worded: "For a trained scientist to claim, or support anyone else's claim, to have transmuted elements is difficult for us to believe and is no more acceptable than to claim to have invented a gravity shield, revived the dead, or to be mining green cheese on the moon."[39] The petition did not succeed, but a few years later Bockris was denied the use of university facilities for a small workshop he was organizing on low-energy nuclear reactions. The investigation of CF effects like anomalous tritium was one thing, but nuclear transmutation was another story. LENR claims were making it even more difficult to advance the case for more benign CF effects like excess heat and nuclear ash.

Transmutation data, like CF in light water, present new conceptual and practical difficulties for researchers working with what is increasingly being understood as merely "conventional" cold fusion (based on Fleischmann-and-Pons-type experiments). Not all experiments producing excess heat produce evidence of transmutations, and while the excess tritium and helium–4 found by researchers like Miles and Claytor at least hold the promise of reconciliation with the expectations of nuclear theory (via the reaction branch of D + D → 4He, for instance), the isotope shifting reported by Mizuno and others is far more problematic. Thus while researchers were making headway in determining conditions for collectively recognizing cold fusion in Fleischmann-and-Pons-type experiments, the new light-water and transmutation data work to destabilize this accomplishment by opening the possibility, for instance, that an experiment producing transmutation effects but no excess heat could be indicative of cold fusion.

Ultimately the new effects associated with cold fusion have produced ambiguous effects in the organization of CF research. Many researchers have tended to shy away from transmutation data until there are more positive results (or until those results can be tied more plausibly to the excess heat and anomalous nuclear ash of conventional cold fusion). But a significant number of other researchers take an opposite tack, arguing that the community needs to remain open to new approaches and explanations. Some of these researchers draw on the new claims to argue that cold fusion can now be seen as an even more revolutionary discovery than people imagined in 1989: "Heavy-water is the orthodoxy. . . . it is true that many of us in the field have followed an evolution or devolution in our thinking. The first shock was Mills, that was a deep heresy. That worried me big time, because I thought my god 'somebody is claiming to see excess in light water.' . . . then onward we went into '92 and other things started coming out, namely transmutations. My god, heavy element transmutations; first it was just potassium to calcium, then it was rubidium, strontium. Now we are seeing more and more and more."[40] Researchers with this point of view have a broader perspective on what sorts of experiments count as indicative of the same kind of phenomenon. Thus research on nuclear transmutation and light-water effects have served to link conventional Fleischmann-and-Pons-type cold fusion with a wider and more diverse population of "new energy" researchers. In this context, cold fusion has become increasingly associated with other kinds of anomalous, problematic, and/or illegitimate energy phenomena from space and zero-point energy to antigravity, Tesla machines, perpetual-motion devices, and nuclear transmutation in biological systems. As I will show in the next chapter, this context has the effect of destabilizing collective and authoritative understandings of cold fusion while at the same time providing an important audience and resource base for continuing cold fusion work.

## *Developments in Theory*

In addition to the instabilities in collective understanding prompted by light-water and transmutation data, an ongoing source of disagreement and divergence occurs with respect to the many attempts to theorize cold fusion. Unlike experimental work, where lack of material resources, labor, and skill constrains the degree to which experiments can diverge from the norm of Fleischmann-and-Pons-style cold fusion, theoretical work requires only access to collectively understandable conceptual resources. These have proved to be more readily available in the post-closure environment, an environment of greater theoretical tolerance and pluralism than was experienced during the controversy. While many theorists insist on working within the boundaries of conventional quantum mechanics in their attempts to theorize CF effects, the illegitimacy of the phenomenon has in an important sense freed up other theorists to make use of more esoteric and radical theoretical resources. While this leads to more accusations of pathological ad hockery from skeptics, amongst CF researchers novel theories are appreciated—that is, as long as they are presented in terms that other researchers (especially those doing experiments) can understand.

Indeed, this remains the problem of theorizing that attempts to either reconcile cold fusion with quantum mechanics and nuclear theory or more boldly undertake wholesale theoretical reconstructions. No theoretical framework has managed to account for all the reported CF effects, and there is no discernible general agreement on which theoretical approaches are even promising. In this sense, post-closure cold fusion is characterized by the proliferation of "pet" theorizing that makes use of novel conceptual (if not mathematical) innovations few CF experimenters understand and few CF theorists agree on.

Rather than surveying the range of these pet theoretical perspectives, I will instead focus on the greatest amount of conceptual overlap as it appears to the experimenters I have talked to as well as some theorists (when I asked them to compare their theories with the theories of others). It seems that this overlap occurs less in the context of full-scale theories than with mid-range theorizing or modeling of what might be going on in the experiments themselves. The general assumption of most CF researchers is not that recognition of the phenomena should force a wholesale revision of nuclear physics and quantum mechanics but that cold fusion constitutes a special case of nuclear theory applied to the conditions of highly loaded Pd-D systems.

CF researchers criticized scientists in 1989 for extrapolating from theories derived from nuclear behaviors in plasmas to nuclear behaviors in solid states. The principles of conventional nuclear fusion research, they argued, depend on pure, isolated, and controlled deuterium environments (such as the super-heated plasmas in magnetic confinement in Tokomak fusion reactors). Cold fusion, on the other hand, is presumed to work in an entirely different manner such that deuterium is able to fuse only under conditions of interaction with the cathode material. This reasoning gives rise to the assumption that highly loaded Pd-D constitutes a special kind of nuclear-active state of matter under conditions that involve impurities, multibody interactions, and a certain degree of inherent unpredictability (and uncontrollability), so that some researchers now believe cold fusion will never be a viable commercial energy source.

Without going into too much detail, the basic argument is that as the palladium interacts with deuterium, it changes properties as a metal-hydride system. This view has been well established in materials science, where metal hydrides are characterized as having an alpha phase and a beta phase. Pd-D moves into the beta phase as hydrogen (or deuterium) loading increases and the deuterium ions take on distinct positions in the metal lattice. This can be easily measured since, for instance, beta Pd is less electrically conductive than alpha Pd; it also has lower magnetic susceptibility and is physically more brittle (which some argue may be the reason for all the cracking in cathodes during CF experiments). Based on this understanding, some researchers (Martin Fleischmann especially) speculate that there is a third gamma phase at very high loadings in which deuterium in Pd can become nuclear active. This view has some support in that the upper limit for beta Pd is considered to be a one-to-one D:Pd ratio and few researchers have reported loading ratios higher than this. Another possibility is that changes occur not to the system as a whole but only to isolated pockets of the metal, around impurities or crack sites and usually at the surface (which has a different chemistry from that of the interior). Edmund Storms summarizes this view as follows: "When the electrolytic method is used, AE [anomalous energy] is produced on the surface within small, isolated regions. These regions heat up in a random fashion, thereby losing the required deuterium content. Thousands of such local regions flash on and off, adding their resulting energy to the total. This process is self-regulating and usually produces a smooth generation of heat energy.

Occasionally, a region will generate enough energy to cause local melting or, once in a while, large bursts of energy are observed. The number of such regions on the surface determines just how much total power will be produced" (Storms 2001, 21).

Even though there is no consensus as to the accuracy of this model, it is a kind of theorizing that gives experimenters some sense of what to expect and how to improve their experiments, as well as a way to draw the varying results of different experiments together. The model advises experimenters, for instance, to avoid using bulk palladium (rods and cubes for example) and instead to maximize surface area by using thin sheets, coiled wires, or in more recent experiments fine palladium powder (known as palladium black) that has been electrolytically deposited on an inert metal (like platinum) in a thin film. Experiments using these sorts of cathodes generate CF effects more consistently than experiments that use bulk palladium samples.

The model does not explain, however, how fusion is possible if only small regions of the Pd-D system become nuclear active, given problems generated by conventional physical phenomena like the coulomb barrier. Somehow the deuterons in these nuclear-active states must get enough energy to overcome the barrier, which normally keeps them apart. If a theorist wishes to work in accordance with basic laws of thermodynamics, then the extra energy must come from somewhere. On this point cold fusion theories move off in all sorts of directions and become increasingly complicated. Generally speaking, however, CF researchers I have talked with tend to favor theories that make the fewest or most minor alterations to conventional theories with which they are already familiar. This is a feature of scientific theorizing in general and is part of an overall wish to integrate cold fusion into existing theoretical structures rather than proposing wholesale revisions (although one can find plenty of theorists who take the second route).

## *Identity and Experimental Practice*

There is now an extensive literature in science studies on experimental practice.[41] While positions vary as to precisely where agency resides in accounting for the production of experimental knowledge, there is general agreement that knowledge develops through some accommodation of the materially embedded experience of working in the laboratory with broader structures of socially embedded beliefs and expectations. Following from this insight, material and social worlds are sometimes thought to be indistinguishable in the context of experimental practice, or if not indistinguishable then at least "mangled" to the point where one cannot be said to determine the other

(Pickering 1995). One way to think about this is to view experimenters as embedded in specific social and material contexts that mediate traditional epistemological idealism around what it means to replicate an experiment. The form of the experimental protocol and the line of investigation experimenters choose depend on available skills, resources, and expectations, and it is much the same for scientists working in routine sciences or on cold fusion (Knorr-Cetina 1979; Galison 1987). As Hacking (1983) has ably argued, experimentation in science amounts to much more than simply hypothesis testing, and this is especially true in cases where there is no general agreement about what hypothesis is being tested.

The case of post-closure cold fusion provides a suggestive new twist on this perspective. The object that engages the attention of cold fusion researchers does not exist as far as mainstream scientific culture is concerned. As a consequence, the normal social and material resources for adjudicating experimental claims both by researchers themselves and with respect to their peers are limited. How does one tell the difference between a good pathological experiment and a bad one when the results of neither experiment are considered legitimate? Under such conditions it may be presumed that anything goes so long as it supports the belief in cold fusion. But this is where it becomes important to understand the nature of CF research as a materially embedded practice. I have tried to show that experiments do matter; one cannot simply believe anything in the context of CF research. The sociality of cold fusion research needs to be distinguished immediately from discussions of the vicissitudes of belief (and/or cognitive dissonance) that accompany research on doomsday cults, for instance.[42] CF researchers as a group cannot believe whatever they want but are tied directly to the structures and contingencies of the experimental practice they engage in.

In the face of these structures and contingencies, CF researchers rely on two things in order to continue their work. The first is their ability to produce an experience of the phenomenon in the lab; they must have a way of "seeing" or experiencing the anomalousness of the data they collect. The second is the need for this experience to somehow transcend the laboratory and be recognized by others. The first condition is satisfied for a large heterogeneous group by configuring cold fusion as an entity that is relatively tolerant of variations in the effects it produces. Excess heat, tritium, helium, neutrons, X-rays, isotope shifts, and more can all be used as independent evidence for cold fusion by any experimenter. Personal experiences of anomaly linked to the configuration of cold fusion from 1989 are important. Above all, the experience of seeing CF effects provides the experimenter with a reason to continue research. As one researcher once told me at a cold fusion conference,

"You only need to see it once to be hooked." Rather than invest a great deal of time and money in a wide-ranging effort to test multiple parameters that might affect reproducibility, many researchers concentrate on generating a robust effect that appears to be truly anomalous and not an instrumental error of some kind. In this way, the researcher develops a relationship to the phenomenon and is better able to extend interpretive flexibility to the results of others. "Seeing it" in this sense provides the basis for participating in a collective effort even though there is no agreement as to what "it" is. Belief in cold fusion plays an extremely important role; there must be a personal reason for scientists to risk their reputations and expend valuable resources, and reports of the experiences of others are often not enough.

The need to move outside the laboratory, however, speaks to the importance of participation in a social world of cold fusion researchers and the various mechanisms that go into coordinating the activities of that world. Cold fusion cannot exist without the association of researchers that produce it. While some individuals are sustained by the naively realist idea that they may invent a demonstration device that will save the day for cold fusion, there is a certain symmetry between the negative experience of researchers so far and literature in social studies of experiment. Demonstrating cold fusion is ultimately a matter of collective action and coordination across social and material worlds, and before any experiment or device can be viewed as demonstrating cold fusion, there needs to be some common understanding of just what cold fusion is. Part of this process involves generating coherence and consistency, indeed convergence, with respect to trajectories of experimental practice. In this case, individual experiences of cold fusion effects become liabilities for the group faced with reconciling and assimilating an increasingly divergent array of data. Excess heat, tritium, and helium–4 are the tip of the iceberg. Aside from isotope shifts and CF effects in light water, there are claims of cold fusion in biological systems and the possibility of a parapsychological connection that links the production of CF effects to the psychic states of the researcher.

The trajectories of post-closure experimental research trace a kind of oppositional tension. On the one hand, researchers work to qualify, isolate, and reduce variability by establishing working criteria for recognizing cold fusion effects. The use of flow calorimetry and the measurement of deuterium loading are examples of this. On the other hand, the same epistemic tolerance that has allowed for the varied effects of heat and nuclear ash has also resulted in an extension of the phenomenon into even less scientifically acceptable territory. CF effects in light water and isotope shifts are examples of this. Thus cold fusion becomes a kind of boundary object (Star and Griesemer 1989;

Fujimura 1992). Through the identity work at cold fusion conferences and the accumulation and communication of experimental data, the phenomenon becomes flexible enough to sustain interest among diverse researchers yet robust and stable enough for those researchers to recognize that they share a common object of study.

The problem is that maintaining the flexibility of cold fusion as a boundary object after closure is a risky business. In a recent book published by the new-age writer Jeanne Manning (1996), cold fusion appears alongside perpetual-motion machines and "space energy" in the context of a developing transcendence in human consciousness. This kind of association can make life even more difficult for CF researchers working in mainstream institutions (consider the case of John Bockris at Texas A&M discussed in chapter 4), and it becomes a threat to attempts at regaining scientific legitimacy. At the same time, however, new associations bring the possibility of new resources for doing research. This development forms the topic of the next chapter.

# *Chapter* 6 Tales from the Crypt

## *Shades of Cold Fusion*

Here is an excerpt from a transcript of ABC television's *Good Morning America* program that aired on February 7, 1996:

> Michael Guillen: It's a device that its inventor says produces a hundred times more energy than it consumes. Now let me say right off the bat that lots of ideas come across my desk that claim to be the energy source of the future, but this one is different. For one thing, the inventor has a distinguished track record. Second, the invention itself has been issued a patent by the U.S. Patent Office. Furthermore, and this is key, independent scientists now claim to have reproduced the results, and major corporations like Motorola are taking a serious interest in it. So, is this potentially the greatest discovery since electricity? Since fire? Good question! Have a look.

In the story that followed, ABC science editor Michael Guillen interviewed an elderly man, James Patterson, a retired chemist from Dow Chemical, who talked of a device for heating water. Patterson, an expert in the manufacture of tiny polymer beads, had developed a technique for coating these beads with micro-thin layers of copper, nickel, and palladium. He placed thousands of beads in a chamber, ran a solution of $H_2O$ and lithium sulfate through them, and applied a current. The result was excess heat. At the time of the interview Patterson claimed a 200-watt output power for every watt of applied power. This excess power effect had been confirmed by scientists at the Universities of Missouri and Illinois using Patterson's beads, and U.S. patents had been granted for the coating technique as well as for the power generation process. The device, known originally as the Patterson power cell (PPC), was now on its way to commercialization through a company called Clean Energy Technologies Inc. (CETI), which was owned by Patterson's grandson, James Redding. Even representatives from the Motorola Corporation had apparently seen the device at work and were interested in Patterson's invention.

Was this cold fusion? Was it the same phenomenon resurrected under a different form? Not if Patterson had anything to do with it; he was wary of the stigma attached to the label and wished to avoid any negative connotation by stressing the differences between his work and that of Fleischmann and Pons. He used regular light water, not heavy water, and his bead apparatus was different from their electrolytic cells. In addition, there was (at least at first) no direct evidence of any nuclear reaction. As far as Patterson was concerned, a conventional chemical explanation would do just fine. Nevertheless, in the ABC news story cold fusion made an appearance close to the end when the host, Charlie Gibson, asked Guillen, "Michael, this sounds like going back to 19 . . . what? 1989."

> CHARLIE: This sounds like the cold fusion debate again.
>
> GUILLEN: Yeah. Remember the University of Utah, the whole cold fusion thing? Superficially this looks like cold fusion, in the sense that you have electricity passing through an electrode that is immersed in salt water. But there are essential technical differences. First of all the beads make this cell absolutely unique. That wasn't like the original cold fusion device. The other thing is that the original cold fusion device used heavy water; this uses ordinary water. So, it remains to be seen whether this is just a variation of the old cold fusion experiment or whether this is genuinely a new phenomenon.

In the evening a follow-up story aired on ABC's *Nightline* with excerpts from the morning show plus interviews with two scientists, Michael McKubre and John Huizenga. Huizenga asserted that Patterson's device was indeed related to cold fusion as a kind of pathological science: "Let me simply say that since Pons and Fleischmann's results were shown to be flawed, there have arisen a whole array of exotic phenomena, including the synthesis of precious metals like gold, which would, of course, be the alchemist's dream, and the light water cells . . . that Mr. Patterson is working on have all been shown not to be producing excess reaction products in the past, and I don't think these people have looked for the reaction products either." McKubre politely disagreed with Huizenga, arguing that cold fusion researchers had indeed been accounting for errors and improving their results over several years and that Patterson's work deserved closer investigation because of its apparent similarities to cold fusion experiments. While McKubre was careful not to embrace Patterson's device as a confirmation of cold fusion, he stated that experts in cold fusion research would be looking into the matter. Rejected by Huizenga and welcomed by McKubre, the Patterson power cell quickly became a focus of attention amongst CF researchers.

Although it looked as if the controversy might be rejoined in a public debate over Patterson's invention, other media outlets did not pick up the Patterson story and scientists were not pulled into the fray as they were in 1989. While CF researchers were unhappy about this, Patterson and CETI seemed to appreciate the lack of attention, arguing that it allowed them to get on with the business of commercializing the technology. Except for the brief appearance of McKubre, the story of cold fusion as it was configured in the Patterson news item jumped from 1989 to 1995 with little mention of developments in between. This temporal disjunction was reinforced by Patterson, who claimed that his discovery was different, and by Huizenga, who wanted to argue that nothing had changed since 1989. At this public level, especially since McKubre was not particularly visible in 1989, Patterson's association with cold fusion made the phenomenon look more pathological than ever. (How, after all, could a retired chemist working alone in his garage in Florida discover the secret of cold fusion when so many scientists working in state of the art laboratories had failed?) It was like a scene out of *The Saint* (see chapter 4). In 1995, cold fusion still did not exist and in the context of the ABC news story the lid on the coffin became a little tighter.[1]

Unbeknownst to the *Nightline* audience, however, McKubre's presence, along with that of George Miley (a nuclear engineer at the University of Illinois who claimed to have confirmed Patterson's results), tied the Patterson device to the underground network of cold fusion researchers who were exchanging up-to-date information through e-mail, websites, and newsletters. McKubre, as we have seen, was already well known for his work reproducing excess heat at the Stanford Research Institute, and Miley, who had conducted a number of his own cold fusion experiments, was also the editor of the *Journal of Fusion Technology*, which regularly featured peer-reviewed articles on cold fusion. Without necessarily wanting it, Patterson now had a cold fusion pedigree, and CF researchers immediately recognized the possibility of a nearly radiationless nuclear reaction in his device. CF effects had been observed in light-water electrolytic cells using nickel cathodes since 1992, and many researchers were already doing experiments using thin films of palladium or nickel rather than bulk metal for their cathodes. That Patterson's beads would work so well made some sense, at least in terms of maximizing the surface area of metal exposed to the hydrogen and/or deuterium in the water.

Despite the fact that CETI's marketing strategy de-emphasized any relation between Patterson's process and cold fusion, the company set up a demonstration experiment at ICCF–5 in Monte Carlo in April 1995. The heat balance of the Patterson power cell was measured with a flow calorimeter designed by Dennis Cravens, an independent CF researcher from Texas who was known for the high quality of his calorimetry. Larger versions of the device (using more beads and running at higher temperatures) were also demonstrated by Miley at the sixteenth biannual Symposium on Fusion Engineering (SOFE) in October 1995 and again by Cravens at the Americas Power Industry Trade Show (Power-Gen) under the CETI banner "New Hydrogen Energy: Feel the Heat." Fleischmann and Pons, working in their Technova-sponsored lab, were supposed to produce such a demonstration, but they had not done so. Here it seemed was definitive proof of cold fusion in the form of a demonstration that heated water, where visitors could "see" the measurements of excess heat for themselves. A few months later CETI even distributed a special Patterson Power Cell Research Kit code-named RIFEX (*R*eaction *i*n a *F*ilm *E*xcited complex) for researchers to work with.[2] By 1996 CETI representatives were also claiming to have observed transmutation effects in their power cells, and they began pursuing contracts to develop their technology for nuclear waste remediation. Since then, however, very little has happened; CETI has been said to be pursuing development of its technology, but representatives have not been collaborating with cold fusion researchers. Jim Redding died in 2000, and it remains unclear what will happen with Patterson's invention.

## *A Technoscientific Ghost Story*

I have made short work of what is really a much longer and more complex story, but the Patterson power cell provides an important glimpse at the broader organization and tensions of cold fusion research as undead science. The story of the Patterson power cell is one of many kinds of technoscientific ghost story. Patterson's invention is a strange if not unworldly kind of object that exists neither inside nor outside science. First, Patterson is a trained chemical engineer known for developing the sputtering technique he used to create thin uniform coats of metal on polymer beads, but he is retired and he built his device in a workshop at his own home, not the usual kind of space where scientific work of this sort is done. Second, without the support of a large laboratory, Patterson did not have the resources to credibly test and measure the reaction products produced in his cells, but this was achieved through a relationship with George Miley and other scientists working in normal scientific institutions. Third, the data collected from Patterson's cells were destined not for publication in scientific journals but rather for the commercialization of a new energy technology. This would not be unusual except for the fact that the basis for the excess-heat and transmutation effects produced by the device has never been recognized by the mainstream scientific community, in part because the attempt to distinguish the excess-heat effect of Patterson's device from cold fusion ultimately failed. Fourth and finally, because Patterson's device became associated with cold fusion, it joined the class of illegitimate and pathological phenomena of which cold fusion is a member.[3]

Patterson and his device, like other facets of the cold fusion research world, are ghostly figures in the sense that they are difficult for the science-studies analyst to locate and difficult to reconcile with an epistemological framework that understands scientific claims to be (or on their way to being) true or false, alive or dead. In the relatively resource rich world of the cold fusion controversy of 1989, Patterson's claims would likely have received scant attention in the face of more detailed and comprehensive studies originating in normal scientific institutions. At the same time, if 1989–1990 did indeed mark the end of cold fusion, then Huizenga's comment on the ABC *Nightline* report would have been the last word on the matter. Instead, despite all the difficulties in producing a salable product, Patterson's cells find life within a post-closure research world that can supply some scientific legitimacy, an ongoing interest, and investors for a company that in return may vindicate cold fusion in the form of a device that can heat people's homes.

In the last chapter I discussed shifts in experimental efforts of the core group aimed at stabilizing an understanding of how to produce cold fusion effects, and this, I argued, coincided with the development of a collective identity for scientists as cold fusion researchers. Collective identification and experimental practice mutually inform each other to produce both stable understandings of the phenomenon and a resource base that enables further research to continue. In this chapter I add another layer to this picture by looking at cold fusion as the focal point for an even more diverse set of social and material actors than I have described thus far. Patterson is not a core-group member; his work does not immediately fit into the trajectories of experimental research I described in the last chapter. He neither presents papers at cold fusion conferences nor advises others on how they might improve their own experiments. While he has collaborated with CF researchers like Miley, they are bound by nondisclosure agreements and so are severely limited in terms of what they can report. Yet even though more secretive than Fleischmann and Pons had ever been, Patterson and CETI, along with many other actors like them, are parts of the post-closure cold fusion research world. How can we account for such strange technoscientific creations as the Patterson power cell?

## Ghost Worlds

Individuals like Patterson and Cravens and companies like CETI along with their investors seemingly take us away from the main activities of the core group of CF researchers. These actors are still doing science but in home workshops and private startups, without doctoral degrees in chemistry, engineering, or physics, and without an expectation of peer-reviewed dissemination of their work, and thus it does not look like the kind of activity I described in the chapter 5. The alignment of identity and experimental practice does not seem to be at work in the same sort of way, and yet for some core-group members, like Miley, Patterson and his device have been important new resources for their research. While skeptical boundary workers like Park configure Patterson as the fringe of the fringe of cold fusion research (an abomination worse than Pons and Fleischmann because now cold fusion in the guise of the Patterson power cell is for sale), CF researchers themselves view Patterson's work as a potential replication, a fresh opportunity, or an intriguing curiosity.

Patterson's presence suggests that understanding the post-closure life of cold fusion only in terms of the activities of the core group is incomplete. More importantly, Patterson's interaction with Miley (amongst others) suggests that our understanding of the dynamics of the core group itself is incomplete. In this respect, we can draw on social worlds/arenas theories in science studies to help complete the picture (Clarke 1990, 1991; Fujimura 1987). The core group is constituted and operates with the context of a more diffuse and associational social world, which in turn occupies a position within the larger scientific arena. As Clarke argues, following Anselm Strauss, "Social worlds are the principal affiliative mechanisms through which people organize social life. A social world is an interactive unit, a 'universe of regularized mutual response,' communication, or discourse; it is not bounded by geography or formal membership 'but by limits of effective communication'" (Clarke 1991, 131).

In the description of mainstream scientific research, the concept of social worlds has been especially useful for capturing aspects of collective practice and organization that cannot be reduced to the activities of formal members and institutions. It is especially sensitive to social and material infrastructures, invisible work, informal communication, and culture in the organization of scientific practice. For this case study, the social worlds concept is useful for capturing the sense of cold fusion researchers as an informally organized, geographically dispersed heterogeneous collective that crosses a variety of micro and macro social locations, only some of which may be immediately recognizable scientific locations. While the concept of social worlds is useful for conceptualizing the relation between insiders and outsiders in science, social worlds are also useful for highlighting the importance of liminal spaces inside science and among the formal members, institutions, and organizations that constitute the primary sites of research on scientific practice.

To sketch out the broader scope of cold fusion research as a social world, let us consider a few of the social contexts that constitute that world.

### NATIONAL CONTEXTS

The media configuration of cold fusion in 1989 made the phenomenon an object of concern in almost all countries with any sort of scientific infrastructure. While the controversy of 1989 took place largely in an American context, post-closure research has in fact been more prominent in countries like Japan and Italy, where differences in political and bureaucratic infrastructures have resulted in slightly different trajectories of research (other important national contexts for CF research include Russia, France, China, and India). Despite the pathologization of cold fusion in these countries, researchers have received some significant support from government agencies and private industry. The work in these countries in turn helps sustain research efforts in the United States, as there is a great deal of international communication and collaboration.

### DISCIPLINARY CONTEXTS

CF researchers come from a variety of different technical backgrounds with their own sets of disciplinary commitments. These include electrochemistry, materials science, nuclear engineering, nuclear physics, physical chemistry, chemical engineering, and systems and electrical engineering. While the research programs of some scientists betray their disciplinary commitments, most are more than happy to characterize cold fusion as an interdisciplinary if not trans-disciplinary problem that actively dissolves fundamental epistemic distinctions between the cultures of physical chemistry and nuclear physics.

### INSTITUTIONAL CONTEXTS

Many CF researchers are full-time professional scientists employed at public and private universities, in government laboratories, and in industry, but a very visible element of the social world of cold fusion is a significant number of retired and semiretired scientists and engineers as well as independent researchers. Indeed, inasmuch as cold fusion has been marginalized as a legitimate research problem in normal institutions, the work of these scientific "others" has become increasingly visible.

### INFRASTRUCTURAL CONTEXTS

Becker (1982) has referred to infrastructural contexts in terms of activities of support personnel who make the primary activity of the social world possible. In science worlds these include equipment manufacturers, funding agencies, institutional administration, laboratory technicians, clerical staff, and graduate students. While infrastructural actors are certainly crucial to all normal science worlds, they seem to play a more visible and prominent role in CF research because the lack of formal organizational support means there are very few of them. As a result, post-closure research is dependent on a high degree of volunteerism (equipment donations and loans, unpaid and low-paid labor) and on the few "friendly" administrators, funding agents, journal editors, news reporters, and politicians for infrastructural support.[4]

Across these contexts individuals engage in coordinated activities directed at exploring, demonstrating, proving, and developing the phenomenon of cold fusion, and to the extent that the activities are linked, cold fusion acts as a kind of "boundary object" (Star and Griesemer 1989) that facilitates the social alignment and material coordination of different actors who work to constitute the nature of that object. Cold fusion both defines and is defined by the alignment and coordination of actors that continue to engage it. The stability of cold fusion as an experimental phenomenon thus depends to a large extent on the coherence and organization of the social world that produces it and not just the core group.

Much of the work of normal science can be described in terms of the broader, more heterogeneous organization of social worlds. Why is the concept of social worlds especially useful in this case? What makes the social world of CF research different from other science worlds is its organization as a post-closure social world. All the activities of CF research (experimentation, communication, interaction, etc.) have been structured or conditioned by researchers' collective experience of the end of the controversy of 1989 (the death of cold fusion). It is an experience that, as I have argued already, alters the distribution of institutional resources for doing science, but it is also an experience that, as I will show below, produces group identification amongst researchers as well as orienting core experimental and theoretical approaches to the continued study of the phenomenon.

It is as a consequence of the experience of 1989 that post-closure CF research constitutes a kind of ghostly social world. This ghost world is like other social worlds in science: it is organized and coherent, and it produces knowledge. But unlike other social worlds, its activities remain invisible and are largely immaterial with respect to the mainstream scientific arena of which it is inextricably a part. In this sense, cold fusion research cannot be understood as being outside science but rather as being in between, underneath, or interspersed throughout normal science worlds. I will examine this distinction and its implications for science studies in more detail in the final chapter, and I will use the remainder of this chapter to provide a sense of the shape and texture of the social world of CF research.

## Core Groups and Cold Fusioneers

While there are a number of important relationships one could focus on in the organization of CF research worlds, it is the tension between the putative core group and individuals like James Patterson that best illustrates the ghostly elements of the social world as a whole. As I have argued, scientists working on cold fusion, unlike scientists working in legitimate core groups, tend to be resource deprived. They must work under the condition of scientific closure that marks their research as pathological science. While the few scientists like Fleischmann or McKubre who managed to hold on to some sort of funding after 1990 have been relatively self-sufficient, the majority of core-group members are only part-time CF researchers. The work of core-group members is hampered on several fronts. It is very difficult if not impossible to get public funding for CF research. Without funding, sophisticated materials and equipment are hard to obtain, and researchers often scrounge and borrow what they need. The lack of funding and the illegitimate status of cold fusion have also led to a lack of graduate students and technicians to do the experimental work. As a result, senior scientists do many of the experiments themselves, and often only when they can find the time. Finally, core-group members have difficulty disseminating their research and finding tolerant audiences. Even those articles that are accepted for review often meet with hostile and frustrating criticism.

Cold fusion as undead science does not survive through the efforts of the core group alone, however. Working alongside core-group members and often in collaboration with them are other participants in the CF research world, whom I have called "cold fusioneers." This is a wide-ranging category of actors that include independent and retired scientists and engineers, inventors, science writers and advocates, entrepreneurs and investors, alternative- and new-energy organizations, and "weird" science enthusiasts. Cold fusioneers engage in a variety of activities that enable the post-closure survival of cold fusion. CF experiments may be performed effectively on low budgets in the garages of independent researchers, funding may come from wealthy investors or organized entrepreneurial efforts, and communication may be facilitated by individuals who publish newsletters and websites, or who organize symposia and workshops. What distinguishes cold fusioneers from core-group members is their relative distance and insulation from the effects of scientific closure. Cold fusioneers operate outside normal scientific institutions and expectations and as a result have access to a wider range of means for doing research.

As with the dynamics of core sets during controversies, core-group dynamics are crucial for making sense of the organization and development of post-closure scientific research. In the case of gravity-wave research, Collins (2000) points out that the legitimacy of the core group working on low-visibility gravitational radiation was protected in part through forms of boundary work that excluded work on high-visibility waves. The case of cold fusion is different: the core group sacrifices some claims to scientific legitimacy for the sake of valuable and much needed social and material resources offered by cold fusioneers. This results in a curious exchange relationship marked by strains and tensions that shape the social world as a whole and keep CF research both alive (experiments continue, research is disseminated, funding trickles in) and dead (CF research is viewed as increasingly pathological the more outside resources are utilized) at the same time.

We do have some models for thinking about this kind of relationship. Collins has remarked that where the experiment in question is perceived to be easy to perform, the core set may be "surrounded by a penumbra of experimental work" (Collins, 1981b, 9) that may be difficult for its main members to accept. A good example of such penumbral work can be found in Travis's (1981) study of the controversy over the chemical transfer of learned behavior. In this case, the central experiment involved transferring the brain of one trained organism (a worm or rat) to another untrained organism by feeding or injection. While there existed a core set of researchers working in about fifteen institutions during the main period of the controversy, the perceived simplicity of the experiment meant that many others could become involved, including students in public schools. Much of this penumbral work was organized through publication of a less than serious journal called the *Worm Runners Digest*, and the research did not appear to have much effect on the dynamics of core-set researchers.

One might argue that something similar developed in the cold fusion case right from the start. In the spring of 1989 it was common for media articles to portray Fleischmann and Pons's experiments as being simple to perform (especially relative to conventional hot fusion research). The article in the *Financial Times* quoted in chapter 2, for instance, suggested that "[Fleischmann and Pons] have apparently succeeded in doing in a simple chemistry laboratory what has not yet been achieved by gigantic nuclear research projects costing hundreds of millions a year" (Cookson 1989, A1). The article sets up an attractive contrast between the simplicity and inexpensiveness of the cold fusion experiment and the complexity and costliness of normal fusion research. Other reports remarked that Fleischmann and Pons conducted their experiments with equipment available in any high school laboratory, generating excess energy fairly quickly. Pons was quoted in the *Wall Street Journal* as saying, "If someone really wanted to do it, I expect they could in a couple of weeks." It is easy to see how the media configuration of cold fusion as inexpensive, simple, and relatively quick to produce could have the effect of drawing all sorts of actors and not just professional scientists into cold fusion research.

While there were not many scientists who believed that CF experiments would actually be that simple to perform, at least the idea of the simplicity of the experiments seemed to attract a lot of attention. However difficult the experiments were to perform, they would not be on the order of magnitude of big-science research in hot fusion. Add to this the atmosphere of excitement around the cold fusion claims and the potentially high stakes of producing a successful cold fusion device, and there is plenty of incentive for anyone and everyone with a modicum of skill and interest to become involved. Indeed, as with the memory transfer case, cold fusion became a hot topic for amateur science clubs, magazines, and especially electronic mailing lists and newsgroups. Fleischmann and Pons's experiments were even attempted in some high schools and by undergraduates at some universities, despite warnings of possible high levels of nuclear radiation and the potential for dangerous explosions. Most of these actors seemed to be penumbral in the sense that Collins and Travis describe: they had very little effect on the core-set dynamics that led to closure. In the post-closure environment, however, some of these penumbral actors have established important relationships with the rejected core group of CF researchers that remains. This is not the case with all penumbral actors, as the vast majority of cold fusion enthusiasts left the scene in 1989, once media attention subsided. What remained was a stalwart collection of individuals with long-term interests and resources for pursuing CF research on their own terms. Unlike core-group membership, the numbers of cold fusioneers and their audiences seem to be stable if not increasing. This may certainly be a consequence of cold fusion as a public controversy; closure has produced a ghost world structured by the dynamics of a rejected core group, a penumbral collection of cold fusioneers, and mainstream scientific culture.

## *Laboratory (After)Life*

What is essentially missing from the post-closure rejected core group is the social and material infrastructure routinely available to legitimate science. Without such an infrastructure, scientific research can progress too slowly and become too frustrating for scientists faced with multiple responsibilities to proceed. Even those who do obtain funding and are able to mount sophisticated CF experiments must still manage the illegitimacy of their research in a climate of skepticism and hostility. One consequence of this, as I explained in the last chapter, is for these scientists to lead double lives, remaining quieter and more publicly circumspect about their cold fusion research than they might otherwise desire to be. There is a palpable feeling amongst core-group members that they have been the subjects of unfair repression, as one physicist told me: "I have to say right now I'm rather discouraged. I guess I think that this is really good physics, I think it's important, I think it's really an interesting problem, and I think that the scientific community in this country has just stomped on it—killed it dead. . . . Anybody who works with me on this is in danger of having their career killed. . . . with respect to getting funding for this area there's essentially—there's almost no avenues to proceed to go to get any funding for this work. I continue in this work only because I think it is important, I should be doing it, but it is at enormous personal cost to myself."[5]

In their positions in normal scientific institutions, core-group members remain subject to the debilitating effects of cold fusion's afterlife. They are not outcasts, nor have they lost the right to practice science, but their work is severely circumscribed, and it is they who bear the brunt of closure's effects, often exhibiting a psychological outlook that is far from rosy. One way of getting a better sense of the ways in which post-closure CF research is circumscribed in the context of core-group work is to look more closely at the specific experiences of researchers. The discussion that follows is based on field notes from an ethnographic study of CF research in one laboratory in a large American research university.[6]

### BEGINNINGS

RF is the principal investigator in our little research group. An engineer of some prominence, he was involved in the design of materials for the first nuclear fission reactors and is now in semiretirement. RF's time is split among teaching, doing research on superconducting materials, and consulting for private corporations. In 1989, after he heard Fleischmann and Pons's announcement, RF became interested in cold fusion, although he had not being working on anything directly related at the time; many of his colleagues at other institutions got deeply involved as well. During the summer of 1989 and into 1990, RF was able to run a number of heavy-water electrolysis experiments, but none produced anomalous results.

As the controversy waned, RF retained a keen interest in cold fusion. In an interview I conducted with him in 1996, RF commented, "I was intrigued and sort of impressed by his [Fleischmann's] efforts to try and determine the critical parameters and make it reproducible and I know the most persuasive argument he was using was getting the right electrode and the right operating parameters—that seemed to make sense." While RF's interest remained, his students' interest did not. In 1991, the students who had been working on the original experiments graduated, and no others expressed any interest in participating in the project. In addition, RF's attempts to upgrade his CF research by obtaining public funding failed.

RF's cold fusion research foundered until 1993, when a new set of experiments was initiated after information started coming back from ICCF–3 in Nagoya. It was at this point that I joined RF as an assistant. Two presentations in particular sparked RF's renewed interest. The first was a report of excess heat from Akihiro Takahashi in Japan. The second was the claim that CF effects were being observed in cells using light water and nickel cathodes. RF liked Takahashi's paper because it was easier to follow and seemed to generate more consistent results than Fleischmann and Pons's work. We resolved to copy Takahasi's protocol, but first we would run some light-water–nickel experiments, modeled on those reported by Reiko Notoya at ICCF–3, because they were less expensive and safer to perform. As RF explained during our first design meeting, "So, first time—I thought there were two aspects, one is let's walk before we run, that is with limited funding . . . let's try light water, just to get a handle on the experimental technique. This is all electrolysis at this stage." This was the first sense I had of our peculiar position, and it illustrates the importance of the light-water experiments to the continued survival of CF. In the absence of funding, heavy-water research was simply too expensive to justify. The light-water results offered us a cheaper way to begin to learn about the experimental technique. For me, this was fortunate. Even as a sociologist I did not have to worry about ruining a light-water experiment. For RF this was also an issue; he was not especially skilled in electrochemistry and calorimetry, and he wanted to have the flexibility of trying different cell designs and protocols without wasting money.

BOOTLEG WORK

Notoya's cells used nickel-foil cathodes and a potassium-carbonate solution as the electrolyte, and these materials were readily available in the engineering department. RF decided not to follow Notoya's protocol exactly, opting instead for an open-cell design, which meant that our heat measurements would not be all that accurate (due to gases escaping from the cell, which carry away heat). RF explained that for our experiments it would be enough to measure the difference in temperature in the electrolyte and the air compared to similar experiments with a blank cell using a platinum cathode. The idea was that we did not need to measure absolute temperature or quantify the heat because a significant temperature difference between the nickel cell and the platinum cell would indicate that something interesting was happening. Then we would get more serious.

RF described our experiment as "bootleg work." We set up in an empty corner of a lab used by the graduate students of one of RF's colleagues. The lab was filled with old equipment that no one used anymore. In my field notes there is a comment from one visitor: "Do all the labs look like this? It looks like a storage room." Our apparatus seemed strangely appropriate for the setting. The electrolytic cell was made from a Rubbermaid plastic container bought from a local grocery store, scrounged bits of wire, and nickel foil that RF bought from a chemical supply company. We did, however, have a high-quality potentiostat (low-voltage power supply) and a neutron counter that RF had kept from previous projects. To this we added a pair of high-quality thermistors bought with discretionary funds and eventually an old circular chart recorder, which would give us a graphic display of the temperature differences in the cell and the air. For the experiments RF did in 1989, students had programmed a computer to collect and analyze the temperature data, but neither RF nor I could do this. We were going to go low-tech.

Other than the potentiostat, there was nothing particularly sophisticated about our apparatus or experiment. RF had some palladium samples he was saving to use later, and he was holding in reserve a sample of a special nickel alloy should we see anything interesting with the plain nickel foil. It was tricky to align the cathode and anode using the bits of wire in such a small container, and I was rather proud of myself when we finally got the thing to work. We did some baseline experiments with a platinum cathode, and then we were ready for the nickel. It was a frustrating process (for me at least). RF was busy with teaching and other research, and I was incapable of running things myself. We worked only sporadically, and always in the late afternoon or evening. If we were lucky, we would be able to run three or four experiments in a month.

It was in the context of this bootleg work that my position as a sociologist was seen as useful. Students in the engineering department had neither the time nor the inclination to work on RF's cold fusion experiments, but as a reasonably skilled, willing outsider, I performed literature searches, monitored experiments, and generally participated as needed. There was a limit, of course. I never started an experiment by myself, and operating the potentiostat even to change the current was cause for consternation. I excelled, however, in reading the digital thermometer and writing down the temperature during the hours that RF could not be present. (My graphic plots of the data were another story however.) In terms of collecting data and running the experiments, the important issue was not so much skill as labor time. As one engineer described the dilemma at his own institution to me, "See you need hands to put anything together, it takes a lot of work to even put together pipes much less instrumentation, check it out, get it working, verify that it's not fraudulent, instruments and etc. . . . The bigger expense is people time, in research—since we have most of the instruments here on site, we don't have to buy very much to put together a system to measure. The big expense is people and people are expensive, and people here are very expensive typically."[7]

In more sophisticated experiments, such as those performed by McKubre at SRI, skilled labor is very costly, and it is nearly impossible to work without funding or the good will of volunteers. For our more modest experiments, my skill level was sufficient to get the job done. Thus I became a ghostly figure in my own right and a kind of cold fusioneer. While I would not be recognized as a legitimate scientist (or even a legitimate science student), in the wake of closure my labor became an appreciated resource in facilitating the survival of CF research (at least in the context of our little group).[8]

Besides collecting data during experiments, I also took on the task of reporting to RF various new developments in the CF research community. This was more of an exchange relationship. RF did not have the time to attend conferences and read all the new papers, and I was doing this routinely for my dissertation research. I used my reports to RF as an opportunity to ask technical questions and get his sense of what he thought was and wasn't a good cold fusion experiment. Like other core-group members, RF was skeptical of non-peer-reviewed reports from individuals without bona fide institutional credentials. At one point I told him about reports of excess energy from a kind of hydrosonic pump device invented by James Griggs, an engineer from Georgia (Griggs 1994). RF asked, "Georgia Tech?" and I said no, he was an independent inventor. RF was visibly unimpressed. "That thing is very doubtful," he told me.

FINDING A NICHE

Over the period of a few months, we ran several light-water experiments without noticing anything interesting. Things changed considerably when KD joined our group. As RF told me, "I didn't do much until I talked to [KD]. He brought up this concept of doing the infrared scanning and I was very interested because I knew the literature, who was doing what, and no one else that I knew of was doing that. So we started and I guess we did two or three experiments just to get the baselines you know and learn how to do it."[9] KD was a technician skilled in the use of the portable infrared scanner. His regular work consisted of using the scanner to analyze heat flow in fission and fusion reactor walls in order to find possible defects and deformities. KD reasoned that the same instrument would be used to scan the cathode during a CF experiment. The scanner would not only quantify temperature fluctuations in and around the cathode, but could also be used to visualize the distribution of heat in the cell. One very important observation would be identifying where the heat was coming from. If the heat originated in the cathode, it would be suggestive of something unusual; if the temperature was higher outside the cathode than inside, it would suggest an external heat source such as recombination or cell resistance.

With KD involved, our commitment stepped up a bit; RF saw an opportunity to do some original research, and the data we obtained were more or less unique.[10] It is important to note that both RF and KD were able to justify their interest in cold fusion to themselves in terms of the infrared investigation of heat flow in nickel and palladium during electrolysis. Even if we did not observe excess heat, RF reasoned that a useful paper could still be written on the subject. At no point did RF and KD frame our work in terms of an organized, concerted research project. This was bootleg science, and we were just puttering around. And yet our experiments produced an anomaly. RF summarized the experience in an interview afterward as follows:

> And the first real experiment was the most frustrating experiment in my life. We thought we saw the system just take off. We were incrementally increasing the current—went up to 1 amp (300 mA/cm2) and then we lowered the current and it wasn't just me but [KD] was watching the potentiostat and we lowered the current to 100, one tenth the current, and normally what happens is the temperature starts going down instead it kept going up for about an hour. We should have continued till it reached a limit, the solution boiling or something but I don't know what happened, either the cell started leaking or something; anyway we went down to a hundred. It was late at night and then we did try to go back, we raised the current again

> and it just refused to do it again and all subsequent experiments using identical conditions as close as I could get it . . . identical samples, same surface finish everything. So all I can say is either something unusual happened, cosmic rays came through or something [laughs] or it was some instrument failure. Somehow if the instrument, the potentiostat wasn't lowering when we thought it was—although it said it was—the meter was registering, but maybe what it said and what it did may have been two different things.

Needless to say, I was as excited as RF was. When, during our interview, I asked how we would know whether the anomaly we detected was indicative of cold fusion, RF replied, "I want to make it clear, I'm just sitting on the fence, there could be something in it, if not cold fusion, maybe something different or interesting, or there may be nothing to it, it's intriguing enough that it's worth looking into it."

This was an important moment in our discussion. Previous to this, we had simply been going through the literature and various protocols for light-water experiments, working through mundane problems that might be encountered in any bench-top laboratory experiment. The anomalous finding sparked a new kind of interest for RF, but his comment to me at this point marked a break in our rapport. We were in an interview situation and not working together in the lab. I had misstepped by mentioning cold fusion in the context of our anomalous experiment, and this prompted the kind of statement one might give to a member of the press. Our discussion quickly returned to normal, but I could make nothing of the break in rapport until much later.

## MANAGING IDENTITY

At one point during an experiment we were running one evening, the dean of engineering walked into the laboratory with two colleagues who were doing a "space check" (a visual inventory of laboratory space in the department). The dean (who had a master key) entered the lab at the opposite end of the room from our experiment, and I was surprised to notice RF immediately go over to greet him. The dean explained what he was doing and started to walk across the room, stopping in front of various old pieces of equipment and asking, "Does anybody use this?" Just RF and I were in the room, and we couldn't answer because we were only borrowing space in the lab from one of RF's colleagues. Eventually the dean came up to the experiment, and RF introduced me; the dean and I shook hands.

I noticed that something was different. Usually, when someone came to see the experiment (they were invited by RF to have a look), RF introduced me as a sociologist eager to get my hands dirty doing science. But on this occasion,

RF said nothing else about me and he didn't offer any explanation of the experiment, even though the dean was looking directly at it. Instead, RF chatted with the dean about the funding for one of his graduate students to do plasma fusion research. After the dean left, RF turned to me and said, "I couldn't decide whether to tell him what we are doing or not. . . . I decided I would like to stay friends." Then he laughed. The occasion with the dean was instructive. As a respected member of the faculty, RF occupies a position of legitimacy that he chooses not to threaten by discussing his work on cold fusion. The threat is not particularly serious; in this instance RF was fairly confident that his relationship with the dean would not be compromised, but all things being equal, it was better to remain silent.

This silence with the dean, however, was not maintained in the presence of other CF researchers. An important part of our activities as a group involved meeting and exchanging ideas with CF researchers, as well as other colleagues and visitors who RF thought would have an interest in our work. At times these encounters felt like a meeting of some secret society. (The pathologization of CF was having an effect on my perception as well.) RF would invite cold fusion researchers to visit the university and give a public colloquium. This would be followed by a private lunch at the faculty club and a visit to the lab or a discussion of our IR-scanner data. Most of the audience at the colloquium had no inkling of our experiments, and often the public colloquium was on a topic unrelated to cold fusion, such as the talk of one scientist from China on the organization of the hot fusion research program in his country. This scientist had a personal interest in cold fusion research and was trying to organize government support for CF work in Beijing. He was especially eager to see our experiment and the possibilities the IR-scanner data had to offer. Sometimes other scientists would be invited to talk about cold fusion. The invitees were always core-group members from reputable institutions, and RF was always careful when introducing a speaker to downplay the idea of nuclear fusion as an explanation for what might be going on.

## LIVING THE SPECTRAL LIFE

As a CF experimenter and a regular faculty member of an engineering department, RF is also a kind of technoscientific ghost; his work on cold fusion remains more or less invisible to the normal legitimate life of the research institution, yet it continues. For RF, on the surface of things at least, there is nothing strange or particularly interesting going on. The experiments we performed were modest, crude, completely inconclusive, and utterly mundane. With respect to our routine of building cells, running experiments, and collecting data, there was absolutely nothing to distinguish our work from the work of the other graduate students in the lab doing more legitimate science. As far as outside observers are concerned, our work is simply good old-fashioned bootleg science, the kind of experiments we might perform out of interest even if the controversy over cold fusion had not taken place. This point can be extended to the more sophisticated and organized experimental programs of core-group members like McKubre, Mizuno, Miley, and others. There is nothing in the day-to-day routine of what they do that marks their work as pathological science—that is, until their work is framed in the context of cold fusion.

Yet normal laboratory life changes with the invocation of cold fusion. As if dredging up some traumatic memory, the words "cold fusion" uttered at the wrong moment would alter our behavior and interaction. This occurred in the context of my interview with RF, but it would also happen when we invited visitors to view our IR-scanner data. If they mentioned cold fusion, RF would become more cautious and circumspect, concerned that we be careful not to overinterpret our data. He would be less circumspect in the company of other CF researchers, who were generally eager to hear about our experiments and how they were progressing, but unlike other kinds of scientific conversations, talk about technical matters would be padded with remarks about the declining fortunes of the field and the negative attention that made continuing research difficult.

In this way the closure of the controversy and the ongoing pathologization of cold fusion affects life in the laboratory. Our research group worked quietly and cautiously with few resources, but the work nevertheless continued. RF's interest unsettles the death of cold fusion and its deployment as pathological science, since RF does the work but for the most part escapes the stigma. In this capacity he is one amongst many others who haunt the stable organization of knowledge, which is predicated on cold fusion's continued erasure. Yet because of this haunting, our work in the lab was idiosyncratic and isolated. Deprived of resources, especially the labor of students, RF and KD felt that promising research had foundered. Beyond the few personal connections of RF to other cold fusion researchers, the quiet low-level work in the lab has no audience. In other laboratories, other CF researchers have the same experience. How is their work able to survive?

## The Work of Cold Fusioneering

At a cold fusion conference in 1993, I ran across a term that well describes part of the life of CF researchers I had been having trouble articulating: "cold fusioneering." While walking through the room that held the poster sessions for the conference, I came across a couple of researchers who were talking about the experiments they were doing in their garages. Trained engineers who did not have permanent employment with any scientific institution, they were giving posters but were not represented in the more widely attended talks at the conference. They spoke of difficulties obtaining equipment, finding inexpensive parts, and running their experiments; they suggested evaluations of others' work but also of their own place in the overall CF research effort. They were doing good old-fashioned bench-top science, "getting back to basics," concentrating on producing simple, high-quality experiments without the need of massive laboratories and expensive equipment. One of the researchers referred to himself and other garage-level researchers as "cold fusioneers."

At the end of the conference, the sociological significance of cold fusioneering became clear. In the last session a number of the most well known scientists were asked to give their assessments of the best CF experiment at the conference. Martin Fleischmann chose the work of Dennis Cravens, who teaches at Vernon Regional Junior College in Texas, saying, "This is real science as it should be done, in his garage on a tight budget . . . producing an unbelievable amount of useful information." One commentary distributed after the conference by e-mail describes Cravens's talk as follows: "A very entertaining presentation on work done in his home lab. Lots of good slides but very few precise numbers were given. The talk was an attempt by Cravens to discuss the good 'recipes' he had found for producing heat. . . . I found myself clapping at the end. Cravens is the first guy I have seen try to do anything with the alleged excess heat."[11]

Cravens not only managed to get his CF apparatus to produce excess heat but also developed a simple calorimetric system for it. He is a technoscientific ghost like RF but is not so much a core-group member as a kind of cold fusioneer. Working outside the normal institutional spaces of science, he remains less subject to the effects of closure; he already works alone and in his spare time, he has never had a large institutional budget to work with, and there is no direct pressure on him to produce competent work. By the same token, there is less need to conform to conventions of scientific presentation or to be careful or circumspect in his analysis, and in this way he is freer to identify as a cold fusion researcher. Indeed, as Fleischmann's comment "This is real science" demonstrates, core-group members see in Cravens the very freedom they desire to have (rhetorically speaking, at least). Cravens and the idea of the cold fusioneer become symbols of resistance to institutionalized science and help to reinforce CF researchers' collective identity as hard-working, skilled scientists dedicated to developing the field in the face of mainstream hostility.

This latter point is a crucial sociological aspect of "cold fusioneering." Cold fusioneers may do experimental work, and they often collaborate with core-group members, but they are also in a position to do a great deal of identity work in being the objects of signification or indeed through their activities in building, mediating, and fostering the infrastructures of the cold fusion research world. This suggests an expansion of the meaning of the original concept of the cold fusioneer. Some cold fusioneers do not do research directly but are active in providing the infrastructural support that helps post-closure cold fusion to survive. Their motivations for doing this are many and varied but to a large extent can be read in the media configuration of cold fusion from 1989. Funding for CF research, for instance, comes in part from entrepreneurial individuals who see a genuine investment opportunity in the development of cold fusion technology. Others support CF research for more ideological reasons, thinking that cold fusion research represents "better" or "more democratic" or "more socially responsible" science. And while there are those who wish to play a part in what they understand to be a significant scientific revolution in progress, there are also those with a more minimalist view that cold fusion is simply "fun" or "interesting" science.

There is ample evidence that while cold fusioneering has become critically important only in the context of the post-closure survival of cold fusion cold, fusioneers have been active since 1989. Here is an account of how one cold fusioneer got involved:

> I had just retired when Pons and Fleischmann made their announcement on cold fusion . . . and having been a student of science all my life I was extremely interested and thrilled when Pons and Fleischmann made their announcement. So much so, it was three o'clock in the morning before I could get to sleep. During that time period I planned a horizon of activities that would be good for at least the next three years and I began immediately afterwards, the very next day, to begin to set up a company . . . and participate in the development of this new technology. . . . My interest was and is one as a systems engineer, not as an electrochemist.[12]

This retired systems engineer became involved in the infrastructural support of cold fusion research immediately after the University of Utah press conference. While he didn't perform any experiments himself until after 1990, he started a newsletter to facilitate the quick exchange of technical information. Initially this work was made difficult by his non-core-set status: "I'd call up a scientist that was working on cold fusion and I'd get a very cool reception. You know like, who are you to waste my time. So the first thing I would do would be tell that scientist I had talked to so and so, and he relayed such and such. . . . this was a whole different situation. . . . here was somebody who was willing to exchange information." After 1990, as news of CF research disappeared from the mainstream and scientific press, the newsletter became an important forum of communication for CF researchers: "Some years after I was told by someone . . . your publication of that newsletter probably saved cold fusion. I won't accept that much credit, but I will accept the credit that it helped." This notion of "helping" often goes beyond the mere mechanics of organizing and publishing newsletters. Cold fusioneers are often advocates who act as conduits of information between core-group members and potential audiences and even as conduits among core-group members themselves.

Jed Rothwell is one such figure. A computer programmer who works from his home, Rothwell provides commentary and articles that are among the first accounts of post-closure cold fusion one would come across in an Internet search on the topic. Rothwell is a contributing editor of the magazine *Infinite Energy* and a regular participant in on-line discussion groups related to cold fusion. Of particular value to some core-group members, however, is the fact that Rothwell speaks Japanese and frequently translates news and information (including technical articles) on cold fusion from Japan. One good example of this is Rothwell's translation of Tadahiko Mizuno's book on cold fusion (Mizuno 1998). The active cold fusioneering of Rothwell and others often opens up channels of communication that not only enable experimental collaboration between researchers but also help to forge the links that constitute collective identity for the group. Perhaps the most obvious context in which to consider this aspect of cold fusioneering more fully is communications.

## The Journal of Fusion Technology

One of the more obvious means of measuring the degree of "liveliness" of a scientific field is looking at the volume and rates of publication of research in journals. After all, peer-reviewed journal articles remain the primary and most valued product of public scientific research. As one might expect given the story I have told so far, the publication rate for articles related to cold fusion went into sharp decline after 1990. In 1989, as we have seen, the Institute for Scientific Information (ISI) identified cold fusion as the topic with the largest number of associated publications out of all the scientific disciplines, but after 1990 cold fusion was nowhere to be seen on the charts of ISI's *Science Watch* newsletter.

The decline in publication rates is reflective of closure processes I discussed in chapters 3 and 4. Rates declined as scientists stopped their experiments and abandoned the controversy in 1990, but in addition the decline reflected decisions by journal editors (such as John Maddox, then editor of *Nature*) to reject or stop reviewing cold fusion articles. While the decline was steep and seems indicative of a quick end to the controversy, it is important to note that the publication rate has never dropped to zero, and a small handful of articles continue to be published in peer-reviewed scientific journals each year. With these data, we see evidence of life after death: a positive rate of publication sustained over a number of years. Yet the rate has been in decline from peaks of over a hundred articles a year in the early 1990s to twenty-five articles in 2000. Data like these have lent support to the claims of the skeptics. Here we do not see life after death, just the dying gasps of the few remaining scientists blindly holding on to their belief in cold fusion. From another point of view, however, the publication data indicate not death or even life after death but rather research settling into a rather normal pattern for a small field. In terms of the overall ecology of science, CF research is simply finding its niche, with a few journals publishing a couple of dozen peer-reviewed papers a year. From this perspective the high rates of the early 90s become the anomaly. The media configuration of cold fusion in 1989 resulted in an unsustainable level of scientific attention that has finally died down, leaving the core group to simply get on with its research. Even though it is too soon to make a judgment about this, the publication rate may support the idea that some normalization of CF research has taken place.

The cold fusion articles that appear tend to be published by a small cluster of specialized journals. The most prominent are the *Journal of Fusion Technology*, the *Journal of Electroanalytical Chemistry*, and the Italian journal *Il Nuovo Cimento*. To a lesser extent articles have also appeared in the *Journal of Physical Chemistry*, *Physics Letters A*, the *International Journal of Hydrogen Energy*, and a number of Japanese and Russian physics, chemistry, and engineering journals. In journals like *Fusion Technology* and *Il Nuovo Cimento*, multiple CF-related articles have appeared over several years, suggesting a neutral if not friendly editorial policy and a set of experts who can act as reviewers for manuscripts when they arrive. In the context of these journals, cold fusion has some institutional status not unlike research on other novel phenomena. Indeed, even though CF articles may be rejected for review by other journals, at least in these few cold fusion finds a stable audience. Or does it?

Until recently, the editor-in-chief of the *Journal of Fusion Technology*, as noted above, was George Miley. While the journal publishes mainly papers related to applied research in conventional nuclear fusion, in 1990 Miley established a regular section for peer-reviewed articles on cold fusion. This section, which featured over a dozen articles a year, became the primary mainstream outlet for the dissemination of experimental work on CF. The situation was problematic from the outset; editors of journals like *Nature* and the American Physical Society journals made the decision to reject CF papers for review, and Miley was one of few editors who actively sought out CF papers to publish: "I stuck with the original decision that papers passing review should appear in *FT* [*Fusion Technology*]. As a result, by default, *FT* virtually 'cornered' the market for CF papers! A backlash quickly followed, with 'hot fusion' members of the *FT* editorial advisory board and some readers vocally questioning my decision. Some declared these papers would 'destroy' the journal" (Miley 2000, 125).

In response to criticisms, Miley made it clear that reviewers would have recognized credentials in the fields of chemistry, nuclear physics, or materials science, and in addition he added a third reviewer for each paper who would be an expert in hot fusion, thus bowing to demands that cold fusion claims be evaluated in terms of the conventions of hot fusion physics. Later, Miley also agreed to stop publishing CF papers that did not deal directly with nuclear phenomena, thus eliminating important topics for the field like electrochemistry and calorimetry (although these articles could still be found in *Il Nuovo Cimento* and the *Journal of Electroanalytical Chemistry*).[13]

In the end, while rejection rates for cold fusion papers were higher than for other articles in the journal, Miley managed to sustain some support for CF research by providing an important mainstream outlet for researchers having difficulty with even having their papers reviewed. And despite constant pressure from the editorial board and accusations that he was irrationally biased in favor of cold fusion, the section survived until 2001, when Miley retired from the editorship after twenty years in the position. While not on the same scale as in the *Journal of Fusion Technology*, cold fusion survives in a few mainstream journals in part because of the influence of supportive editors and reviewers, but this is a precarious situation subject to change should the supportive individuals leave their positions.

Like the case of the Patterson cell and my experience with cold fusion experiments in the laboratory, the saga of George Miley and the *Journal of Fusion Technology* is a kind of ghost story. Cold fusion became visible (annoyingly so in the opinion of many scientists) to normal science worlds through the efforts of Miley, but its presence in the journal seemed unstable and impermanent, especially since the editorial board wanted to see the section disappear. With Miley's retirement, after the backlog of papers has been published, the journal will likely drop its support for CF research, thus contributing to a further decline in publication rates and bolstering evidence for the death of cold fusion. In spite of this, Miley's invisible infrastructural work in keeping CF afloat in a mainstream journal and the scant evidence of life supplied by data on CF publication rates are merely epiphenomena of a deeper, richer, and more stable research world. The presence of cold fusion in places like *Fusion Technology* or *Physics Letters* A (as with the case of Fleischmann and Pons's heat-after-death paper) is indicative of an undead science that operates beyond the limits of our usual analytic vision. While the dissemination of CF research limps along in mainstream journals, there has been a virtual explosion of information in the form of web-based publications, magazines, and newsletters.

## Cold Fusion On-line

Beyond conferences and journal articles, cold fusion research survives through informal and dispersed communication that takes the form of electronic mailing lists, web pages, print newsletters, and magazines. In the post-closure research world, these noninstitutional media have become a critical resource for the transfer of information as well as the maintenance of group identification and solidarity. Although researchers face off-putting hostility and frustration in their attempts to present CF work in the institutionalized publication of mainstream journals and conferences, they find a climate of relative openness and tolerance in the context of websites, newsletters, and magazines. Measuring communication in these alternative media tends to be rather difficult, but once we begin to account for them, cold fusion research begins to look much more lively than an examination of mainstream journal publication rates alone would indicate.

Since 1989 the list of alternative venues for the dissemination of CF research has grown considerably, and there are two basic forms that these may take, self-publication and moderated publication. Self-publication is a relatively old practice in the natural sciences that occurs when scientists wishing to disseminate either preprint material or rejected papers make use of photocopiers and faxes to distribute their ideas. Moderated publication is a more organized communicative practice in which an editor or some other individual mediates the flow of information. Journals are examples of moderated publication, but so are magazines and some managed electronic discussion groups and websites.

As Internet technologies have become more ubiquitous, self-publication has become increasingly widespread. Electronic preprint and document services, e-mail lists and newsgroups, and personal websites have all become means of helping ones' ideas survive in the face of rejection by normal science media. The Internet especially enhances the longevity of ideas by allowing for their distribution across a wider social arena and for more flexible and accessible storage. In 1989, for instance, varied audiences of scientists and the lay public alike gained exposure to media-configured cold fusion claims via the distribution of news items and commentary through electronic newsgroups. After closure and into the mid–1990s, those who continued to make use of these newsgroups were still encountering messages and information from 1989. In this way, conceptions of cold fusion from 1989 tend to be continually reproduced in public electronic spaces despite the shifts that have taken place in experimental practice.

The use of personal websites for publishing information related to CF is similar to the use of newsgroups, except that it is easier for the individual to control the distribution of information. Currently there are dozens of websites operated by individual CF researchers disseminating their own ideas, as well as sites that compile information from numerous sources. In addition there are hundreds of sites that feature descriptions of the cold fusion case (most of which are outdated) and links to the core cold fusion sites. As with Internet communication in general, websites may come and go, but there are a number of sites that have become more or less stable through regular updates of information, connection to off-line organizations, and/or integration with other aspects of the social world of cold fusion. Consider this sample of the kind of information one may encounter on the Internet.

### PLASMA DISCHARGES ON PALLADIUM (HTTP://WWW.NDE.LANL.GOV/CF/TRITWEB.HTM)

Mark Schwab is a Ph.D. student working as a research assistant for Thomas Claytor at the Los Alamos National Laboratory. Schwab created this website, which contains little other than a few abstracts and a scientific paper authored by Claytor, D. D. Jackson, and D. G. Tuggle, "Tritium Production from a Low Voltage Deuterium Discharge on Palladium and Other Metals." The paper is a report of anomalous tritium measurements in a variation of the cold fusion experiment using deuterium plasma discharge (see chapter 5). It is significant in that it is authored by a core-group member, has been praised for the quality of data it presents, and appears only on this website and not in any peer-reviewed journal.

### WEIRD RESEARCH, ANOMALOUS PHYSICS (HTTP://WWW.ESKIMO.COM/~BILLB/WEIRD.HTML)

Bill Beatty is a Seattle-based electrical engineer who in his spare time maintains a website for science hobbyists and "weird" science. An individual with an interest in scientific anomalies, Beatty collects and organizes web-based information (links and articles) on cold fusion as one amongst many other anomalous phenomena in physics (including Telsa coils, antigravity, ball lightning, and magnetic water). Beatty also moderates a number of e-mail lists on topics related to new-energy research. His Vortex-L list has become an important forum for the interaction of cold fusioneers and core-group members as well as curious bystanders:

> The Vortex-L list was originally created for discussions of professional research into fluid vortex/cavitation devices which exhibit anomalous energy effects (ie: the inventions of Schaeffer, Huffman, Griggs, and Potapov among others). Currently it has evolved into a discussion on "taboo" physics reports and research. SKEPTICS BEWARE, the topics wander from Cold Fusion, to reports of excess energy in Free Energy devices, gravity generation and detection, reports of theoretically impossible phenomena, and all sorts of supposedly crackpot claims. Before you subscribe, please see the rules below. This is a public, lightly-moderated smartlist list. There is no charge, but donations towards expenses are recommended.

While core-group participants often speak negatively about the use of electronic lists for the discussion of CF research, forums like Vortex-L attract critical and faithful audiences for core-group work. Discussions on the list rarely influence experimental work in core-group laboratories, but on occasion CF researchers will engage participants directly and indirectly on various topics.

### THE INSTITUTE FOR NEW ENERGY (HTTP://WWW.PADRAK.COM/INE/)

The website describes the Institute for New Energy as "An official US nonprofit technical organization and a co-sponsor of the International Symposium On New Energy (ISNE). It is a membership organization whose monthly newsletter 'The New Energy News' (NEN) reports the latest findings in New Energy research. The Institute's primary purpose is to promote research and educate society of the importance of alternative energy. It is also related to the International Association For New Science, whose goal it is to institute a paradigm shift in science and healing." Through the efforts of Patrick Bailey and Hal Fox, in addition to sponsoring symposia and publishing *New Energy News*, the INE website hosts articles and commentary. Cold fusion research takes on a central role here in a broad, eclectic program of research on "new-energy" technologies. Such technologies, associated as they are with the legacy of perpetual-motion machines and a variety of other rejected claims, do little to enhance the mainstream legitimacy of cold fusion, but the INE does provide new audiences and opportunities for cold fusion work.

### EARTHTECH INTERNATIONAL, INC. (HTTP://WWW.EARTHTECH.ORG/)

Scott Little is a researcher for Earthtech International, Inc., which he describes as "a privately funded research organization dedicated to the exploration of new frontiers in physics. Our activities are primarily centered around investigations into various aspects of the Zero-Point Field. In addition we perform evaluations of reported 'over-unity' energy devices. We specialize in performing accurate power-balance measurements using calorimetry." Earthtech was founded by Harold Puthoff, a theoretical physicist known for his work in the controversial field of zero-point or vacuum energy (a measure of the energy density of empty space). One speculation is that cold fusion effects may be related to zero-point energy in some way, and so Earthtech sponsors small-scale attempts to replicate some CF experiments, performed by Scott Little. While Little is not active in the core group of cold fusion researchers, his efforts have been illustrative of the difficulties in dealing with the experimenter's regress. He has been active in on-line forums related to cold fusion and publishes all his experimental data on the Earthtech website. As a result, some of the most active discussions in Vortex-L, for instance, have been about the strengths and weaknesses of Little's experiments.

These websites represent only a small subset of what can be found on-line. The subset is curious in that it presents cold fusion as a mix of respectable core-group science (the website at Los Alamos), more eclectic energy research (the Institute for New Energy), and the efforts of individuals to sustain a level of public debate and dialogue about cold fusion (William Beatty and Scott Little). The hybridization of these efforts occurs with cross-fertilization (or hyperlinking) among the various websites, creating an image of CF research that differs markedly from the work of core-group members alone. Few core-group scientists are affected or influenced by representations of their work on these websites, but as an open medium on-line communication becomes a source of both opportunity and anxiety for core-group members and cold fusioneers alike. While the web publication of Claytor's tritium results has been useful for researchers (especially those in parts of the world where journal access is difficult), the web publication of data from Little's experiments is more disturbing. Little's replication attempts have all been unsuccessful, and this has had the effect of strengthening the public case against cold fusion. As if conditions for survival were not difficult enough, this puts CF researchers in a position of fighting rearguard actions against negative replications that do not even appear in journals. The on-line discussions of CF experiments are perfect examples of the an extreme version of the experimenter's regress, where there is simply no mechanism (formal or informal) for resolving interminable disputes over experimental competence.

## Infinite Energy *Magazine*

Alongside the open media of web publication and electronic discussion, however, are the moderated media of private newsletters and magazines. The three most prominent are *Fusion Facts*, *Cold Fusion Times*, and *Infinite Energy: Cold Fusion and New Energy Technology*. *Fusion Facts* and *Infinite Energy* publish articles about cold fusion as well as other radical claims related to the production of energy, but much of their material and readership stem from a central focus on cold fusion. The table of contents from one issue of *Infinite Energy*, for instance, lists an article that discusses the technical elements of an important CF experiment performed by researchers at Osaka University.[14] This appears along with articles on modern methods of transmuting mercury to gold, the invention of a new high-efficiency motor, and energy production resulting from elemental transmutation in biological organisms. In addition, each issue also features technical letters from researchers, advertisements for energy-related research equipment, book reviews, and comics.

In general the magazine defiantly promotes the idea of cold fusion and new energy, while at the same time acting as a conduit for technical information in a manner not unlike the identity work that occurs at cold fusion conferences. The July 1999 issue of the magazine, for instance, celebrates the tenth anniversary of the announcement of cold fusion with the title "Cold Fusion Celebrates 10 Years of Revolutionary Science and Technology." It features commentaries from core-group scientists on the progress and prospects of cold fusion research, an article by Martin Fleischmann on the history of his research with Pd/D systems, an article by Edmund Storms entitled "My Life with Cold Fusion as a Reluctant Mistress," and a complex theoretical paper by Elio Conte that offers an explanation of cold fusion based on a reinterpretation of Rutherford's 1920 model of the neutron, using quantum mechanics and a theoretical entity Conte calls "biquarternions." There is also a transcript of a lecture given at MIT by the Nobel laureate Julian Schwinger in 1991. Before his death in 1994, Schwinger took an active interest in cold fusion, resigning from the American Physical Society in protest over what he saw as the censorship of cold fusion in APS publications. Finally, the issue offers a "special report" on attempts by MIT researchers to cover up and misrepresent their negative replication data from 1989. The contents of this issue demonstrate the degree to which core-group scientists have been integrated with the kind of cold fusioneering that the magazine represents.

Although the magazine is written for the most part in a popular and accessible style, many of the contributors and readers of *Infinite Energy* are the same as those who read and publish CF articles in the *Journal of Fusion Technology*, and several prominent CF scientists are on the advisory board of the magazine. In a field where forums for communication are severely constrained, *Infinite Energy* provides an important alternative by supplying new technical information quickly and cheaply. The magazine is published and edited by Eugene Mallove, who, as we have seen, wrote the only popular book in support of cold fusion research (Mallove 1991). Mallove started *Infinite Energy* in 1995 after leaving the editorial post of another cold fusion magazine (*Cold Fusion: The Magazine of the Water-Fuel Age*), and has until recently run the operation from his home at a financial loss. *Infinite Energy* has the highest production value and the widest circulation amongst the CF newsletters. Mallove routinely prints about five thousand copies of each bimonthly issue, which is anywhere from fifty to a hundred pages. The magazine has several thousand subscribers and sells up to 80% of the additional 2,400 newsstand issues it prints.

Mallove himself is something of a self-styled crusader. In 1989, while working as the chief science writer in the public relations office at MIT, he reported on cold fusion for MIT's *Technology Review*. Mallove's experience at MIT during the cold fusion controversy prompted him to quit his job in 1991; he claims that the scientists working on cold fusion at MIT misrepresented their data in an effort to discredit Fleischmann and Pons, and that the MIT administration blocked his attempts to publish more sympathetic articles. Mallove firmly believes that many of the vocal critics of cold fusion have conspired to "kill" it in order to protect their interests in the continuation of federal funding for hot fusion research. His occasional "rants" in editorials and electronic newsgroups are infamous, and an important example of reverse boundary work. In one editorial he writes,

> When Drs. Martin Fleischmann and Stanley Pons made their seminal announcement at the University of Utah on March 23, 1989, no one could have predicted that they were truly heralding the end of the Fossil Fuel Age. . . . It took many months for the reality of the astonishingly powerful and mysterious energy source within water to be confirmed by a host of other laboratories. Now, there are disbelievers only among the under-informed, those who accepted the anti-science propaganda of cold fusion's opponents, and members of the Flat Earth Society—the professional "skeptics" who actively wage war against cold fusion. The latter are devout Believers that there can be nothing new under the Sun in physics, save for what is supposedly

> allowed by physics' venerable "sacred texts." The last six years have been intensely frustrating, as the predicted agonies of a Kuhnian paradigm shift in science have been visited on hundreds, then thousands of cold fusion researchers world-wide. To be sure, enormous progress has been made in sharpening experiments and persuading other scientists to put aside their prejudices, but *the goal posts keep getting shifted*. (Mallove 1995)

Mallove and *Infinite Energy* are also ghostly figures. His editorial can be read at the same time as both the idiosyncratic ranting of a crazed lunatic and an important framing narrative for the organization of cold fusion research after the closure of the controversy. In its style, the editorial reproduces cold fusion research as pathological science, reinforcing the notion that cold fusion is dead. At the same time the editorial articulates a difference that marks CF researchers as "better" scientists than their opponents, thereby justifying the legitimacy of continuing the research that keeps cold fusion alive. Like the cold fusion conferences, *Infinite Energy* becomes a context in which the publicly accountable identities of scientists working in institutions can be relaxed.

The tension that produces the cold fusion conferences and *Infinite Energy* also produces a phenomenon that is recognizable by a critical deuterium content but can also be seen as an extension of new-age alchemy. Researchers' experience with closure and its consequences leads to an organization of practice that makes this duality possible. Talking about the cold fusion conferences, one researcher comments, "I do always find it kind of annoying that you see some sort of fringe activity which is definitely not scientific, para-scientific at these meetings."[15] For this researcher "fringe activity" hurts the legitimacy of the field, and in particular his attempts to revive research in his own institution. When I asked one of the scientists who organized the 1993 cold fusion conference about this, he responded by saying,

> Two arguments weighed in favor of letting pretty much anybody except outright lunatics—even then it's really hard to tell. One, people can't get to conferences very often if they can't present so you have to give them an opportunity to present. The second thing though, in this field, it's very difficult to tell what's good and what's not and I was struck by something that Martin said at the closing, he awarded Dennis Cravens his vote for the best paper of the conference. If we had applied a strict filter we would have never have let him talk cause he comes from a little school and his abstract didn't sound all that interesting. So you have to allow for the possibility that somebody you've never heard of who's working out of his garage might actually know something that you don't know.

And to a question about the cold fusion newsletters he replied, "It's slightly embarrassing. . . . It's going to turn out that 90% of this sort of UFO type stuff is just wrong; it's just a mistake. But 10% might be right, but if one of them is right that's interesting. So I think we need to pursue them and we need to pay attention to them. And not shut them down."[16] In the post-closure world of cold fusion, the "fringe activity" of some researchers is problematic, but it is also a palpable resource.

## *Japanese Ghosts*

My final ghost story presents a different kind of fringe activity. For the most part the arguments of this book have focused on the experience of cold fusion researchers in the United States and to a lesser extent Europe, but until recently Japan has been a major center of CF research in the post-closure period. This became clear in 1992 with the first cold fusion conference in Japan and Fleischmann and Pons's contract with Technova, and throughout the mid–1990s it was obvious that even Fleischmann and Pons's efforts in France were a very small part of a large ongoing research effort in Japan. Unlike the experience in North America and Western Europe, cold fusion research in Japan actually received more infrastructural support for research after 1990 than before.

In the months after Fleischmann and Pons's initial press conference, Japanese interest in cold fusion was extremely high, buoyed perhaps by government and corporate attraction to sources of alternative energy and an extensive institutional apparatus for funding speculative applied-technology projects. Through 1989 and 1990 there were CF research groups operating at many major universities just as elsewhere, but the media and scientific closure of the controversy that centered in the United States also affected public and scientific perception of CF in Japan. Japanese scientists came under pressure from their colleagues to give up their research as claims of cold fusion as pathological science hit the Japanese media. After 1990 CF research in Japan could have moved into a post-closure phase similar to what I have described in the United States, but differences in the organizational culture of R&D funding in Japan (Traweek 1988) meant that certain individuals who were well placed in the organizational hierarchy could help keep CF research projects afloat.

There are other factors at work as well. Despite the strict hierarchies of the various organizations, funding opportunities in Japan are more diversified than in the United States, allowing better access to support for Japanese CF researchers, even in the face of skepticism. The overlapping of corporate and academic science worlds in Japan also changes the kind of boundary work that occurs. Scientists are less likely to define themselves in terms of their discipline than in terms of which project they work with. Lastly, public displays of skepticism and character assassination of the kind found at conferences in North America and Europe have been rare in Japan.

In 1993, as noted above, a joint research program was initiated by the Japanese Ministry of International Trade and Industry (MITI) and the Agency for Natural Resources and Energy. The program was run by the New Energy and Industrial Technology Development Organization (NEDO)[17] and received funding from several major Japanese companies, including MHI, Hitachi, Toshiba, Nippon Steel and NTT. NEDO had a budget of $30 million to spend over four years, and much of this went to establish the Institute of Applied Energy (IAE) and its pilot New Hydrogen Energy (NHE) project, with laboratories in Tokyo and Sapporo. The aim of the NHE project was to "confirm and demonstrate excess heat generation phenomena" and to make "excess heat generation controllable." The NHE project employed close to one hundred researchers, and cooperative links were established among scientists at Japanese universities and industry labs, as well as researchers in the United States (including McKubre, Miles, and a few others).

The stark contrast between the experience of CF researchers in the United States and Japan suggests that the notion of closure discussed in chapter 3 is culturally specific. That is, a controversy that ends in one cultural context may not be over in a different context. To a certain extent this would seem to be the case. At least until recently, Japanese CF researchers have not needed to conduct bootleg experiments with scrounged equipment, nor have they made much use of the Internet or CF newsletters to disseminate their work. The extent to which different organizational cultures and collective identities may help explain the difference in the reception of the cold fusion claims remains to be studied in detail. However, it is worth noting that Japanese CF research, along with CF research in other countries, has not been unaffected by the outcome of the controversy in the United States. The head of the NHE project, Hideo Ikegami, made it clear that part of the strategy of Japanese CF researchers was to avoid the label of cold fusion by adopting the label "new hydrogen energy" for their work. This strategy reflects a self-conscious attempt to avoid the stigma associated with cold fusion while leaving open more conventional avenues of explanation for the phenomenon.

In the global context of post-closure cold fusion research, the NHE project was a very palpable technoscientific ghost. The resources marshaled by the project not only enabled continued CF research in Japan but also allowed for the funding of CF conferences and collaboration with researchers in Europe and the United States. This created a genuine problem for scientists in attributing closure to the case of cold fusion. For many scientists in the United

States, their cultural isolation partly solved this problem. When I mentioned the NHE program to one critic I interviewed in 1996, he replied that he had not heard of the Japanese effort. Other scientists used cultural difference as the explanation, arguing that perhaps the Japanese had been duped by Fleischmann and Pons or that the Japanese are better at applied research than basic science. Alternatively, the Japanese program suffered from an institutional pathology driven by the need for economic success; It would grasp at any possibility, even cold fusion.

While Japanese CF research can be ignored or dismissed by scientists in the United States (or in Japan for that matter), the material effects of the research cannot be. Experiments are performed, results are published, and devices are built. All of these things help sustain cold fusion after the experience of closure in 1989–1990. The importance of the Japanese program was recognized as news circulated that the NHE project was to be shut down. After six years, the project had failed to meet its main goal, the construction of a reliable demonstration device. As one American CF researcher wrote,

> Rumor has it that the NEDO cold fusion program has finally bit the dust. They have reportedly cut off funding to SRI, and they are preparing to shut down the operation in Japan too. Selling off equipment, giving people their walking papers. I have not confirmed this at the highest levels yet, but it is what I have been expecting for some time. As I reported here and in I.E. [*Infinite Energy*], during ICCF6, the NEDO managers told me the program was in deep trouble and they were not sure of additional funding. The fiscal year in Japan ends in April, so now is when the money runs out. (That is also the end of the academic year, by the way.) . . . I guess it is up to us now. If we cannot pull off a good demo, cold fusion will die, maybe for years, maybe forever. Many other nascent technologies have withered on the vine.[18]

The NHE project ended in April 1997. Skeptics like Douglas Morrison argued that the Japanese researchers finally came to their senses, adding further proof that cold fusion was not real. After all, the Japanese had made a concerted effort in spite of the events of 1989, but they tried to replicate the CF effects and failed. Surely there was nothing left for supporters of CF to fall back on. Many CF researchers argue, however, that the NHE project did not pursue the right kind of experiments, was not concentrating on replicating the most robust experiments but rather continuing to work blindly with Fleischmann and Pons's original protocol. Moreover, it had neglected the advice of researchers who were not working with it.

Today the organization of Japanese CF research more closely parallels the situation in the United States. It survives now through experiments performed with equipment left over from the NHE project. The main laboratory in Sapporo was shut down, but equipment and even researchers' time have been borrowed by other groups working, for example, at Hokkaido University (T. Mizuno, R. Notoya, T. Ohmori), Osaka University (A. Takahashi, Y. Arata), and Mitsubishi Heavy Industries (T. Itoh, Y. Iwamura). Japanese CF researchers have also founded their own professional society (JCF) to promote CF research in Japan and provide a forum for collaboration and dissemination of research.

## *Between Science and Oblivion*

There is certainly much more to be said about how post-closure cold fusion research is organized, but I will be content to return now to the problem posed in chapter 1. At this moment, there is a group of researchers actively working on cold fusion. These researchers appear to be experimenting, publishing, collaborating, and going to conferences in much the same way as "normal" scientists do, but there is a twist. The controversy over cold fusion has ended, and CF research survives, but it is in the context of a kind of shadow world of science. Like ghosts, scientists' actions with respect to cold fusion are often unseen and unrecognized by other scientists working in the same institutions. CF research may go on at scientists' homes or at night, and it will not be reported in departmental newsletters or discussed at colloquia. Alternatively, the research may be more visible but not identified as "cold fusion." The scientist might instead be studying "anomalous properties of deuterated metal hydrides."[19]

As ghosts, CF researchers walk a fine line between mainstream scientific legitimacy and access to the unconventional resources available by also being perceived as not doing science. If they travel too far in one direction, they risk losing what status and resources are left to them as scientists (as in the case of John Bockris). If they don't travel far enough, on the other hand, they may not have the material or cultural resources necessary for continuing their research. (More modest experiments are less likely to attract private funds than experiments linked to the production of commercial applications.)

It should now be clear that post-closure research on cold fusion is not limited to one or two stubborn individuals but rather can be characterized as a kind of research world shaped by the participants' experience of closure and pathologization. To the extent that CF researchers are collectively able to add to the corpus of knowledge about the physical world, they must be recognized as doing scientific work. If the task of the sociology of knowledge is to account for the relationship between the production of social organization and certified knowledge, then the activities of the post-closure CF researchers become important elements in that account. In the remaining chapter I will sketch a theoretical and methodological rationale for why this is so.

# *Chapter 7* A Hauntology for the Technoscientific Afterlife

## *Asti*

In the fall of 1998, I attended a conference on cold fusion held in Asti, Italy. This conference was different from the larger International Conferences on Cold Fusion in that participants were able to sustain more detailed conversations in the sessions and at social gatherings. As at previous cold fusion meetings, I had gone as an observer with my tape recorder, but this time I had also brought along the draft of a paper that had begun to develop some of the themes I have been arguing in this book. The paper was then titled "The Undead Science of Cold Fusion: Towards a Hauntology for the Technoscientific Afterlife," a typically pretentious title for a graduate student paper. Early in the conference I handed out a few copies of my paper, hoping to get some feedback on the accuracy of technical matters described in the text. I was in for a surprise.

As it happened, I had been lucky enough to meet an Italian graduate student who worked on cold fusion. We became friends, and not only did she spend a great deal of time explaining to me who was who and what they were working on, but she also acted as an impromptu translator for conversations between sessions in Italian. During lunch on the third day of the conference, my friend informed me that there was some trouble brewing at the next table. I asked what it was. It appeared that one of the Italian scientists had obtained a copy of my paper and was translating the title to his Italian colleagues as "The Dead Science of Cold Fusion." These scientists were being told that I was there to do a hatchet job on cold fusion and their work, and they were getting pretty angry. While there was some joking talk (I think) of "taking care of things" later outside, I decided it would be prudent to intervene, and with the help of my friend I attempted to explain the title of the paper. "Cold fusion is not dead but undead, like a ghost. . . . You know, the soul lives on long past when everyone else considers the person to be gone, but it's invisible, hard to see."

"Cold fusion is real," came the reply, "you cannot deny the experimental evidence now."

"Absolutely," I affirmed; "I am not disputing that, but I'm trying to make another point—a sociological point." Although I kept at it for a few minutes, I don't think I was successful in convincing anyone at the table that my metaphor for understanding cold fusion research was at all apt. Indeed, no one except perhaps (and most appropriately) my graduate student friend found the ghostly identity I was imposing on them at all likable. My friend was just finishing up her thesis and was already encountering problems talking about her cold fusion work with anyone other than her supervisor (who also was a CF researcher). As far as the scientists at the table were concerned, however, cold fusion was very much alive, more so in fact than ever before, and the problem was how to break through the barrier of the few ignorant yet influential scientists who tended to make things difficult.

## *Cold Fusion Wanted: Dead and Alive!*

I began this book by posing a dilemma. How do we account for the simultaneous observation that cold fusion is both dead and alive? It should be clear by now that more people than just historians and sociologists have an interest in this question. I have mobilized the notion of scientific "life" and "death" first and foremost because they are actors' categories that mark the boundaries of legitimate and illegitimate practice in science. Scientific life in this sense does not imply certainty but rather epistemic legitimacy; CF researchers do not yet collectively claim to know what cold fusion is; they claim only that it is a real anomaly of genuine scientific interest. If granted the status of scientific life, cold fusion qualifies as a legitimate experimental phenomenon worthy of the attention and resources of normal scientific institutions. The category of scientific death, on the other hand, carries a large epistemological price tag. If dead, cold fusion passes into the history of failed scientific claims, and joins N-rays, polywater, and ESP as part of the litany of mistaken beliefs. Scientific death in this sense is a corollary of Popperian notions of scientific progress. Science progresses not because certainties are discovered or determined but rather because errors are uncovered and discarded. Under this view the role of scientists is not to give birth to the truth so much as to endeavor to kill off errors. It follows then that the stakes of the question of whether "it" is dead or alive are much higher for the actors than they would ever be for science-studies analysts. At the root of all this actor-generated musing about the mortality of scientific claims, real money, real credibility, and real knowledge are on the line.

Perhaps for this reason, sociologists of scientific knowledge prefer to demur on the practical question of scientific life and death. Asked whether I believe cold fusion is real, I tend to reply that "quite a few scientists definitely think so." Science-studies analysts, unlike our more philosophically inclined cousins, are exhorted to let scientists decide whether a claim is true and whether a phenomenon exists or not. We are not doctors. Not only should we not pronounce on the life or death of our subjects; there are few that would believe us anyway. I introduced the book with this point. Our task is to account for, and report on, how decisions about life and death are made based on what we observe as social scientists who study science. Sociologists of science like Bruno Latour and Harry Collins may disagree on who or what participates in the decision making we observe, but they would agree that the analyst should not intervene. We need to let the actors speak for themselves while restraining our realist and relativist presuppositions in order to listen to what they have to say.[1]

The injunction to shut up and listen to the actors is extremely important advice for some philosophers and boundary workers with an epistemological axe to grind, but there would seem to be a limit to this gesture. Listening to the actors works only where all relevant actors are indeed present (or can be found) so that they can be heard. This is the problem we face in accounting for scientists' decision making about the life and death of claims. Who is a party to the process? How much can we rely on what we normally see and hear? The story of cold fusion prompts us to think more closely about these questions. Clearly this is not an open-and-shut case, as resistance to the death of cold fusion is too visible and obvious. Yet my experience in the field also prevents me from adopting a simple "he said, he said" relativistic position on the matter. It does not appear to me in this case that science is a pluralistic world where every idea may find its niche.

Thinking about these issues has prompted me to tentatively violate the injunction to shut up, moving beyond the actors' categories of life and death in order to occupy a different analytical space for making sense of differences in the visibility and power of actors in technoscientific arenas. In this final chapter, I want to return to the analytical problem that I laid out in the first chapter, informed now by the hindsight of the intervening chapters. What does it mean for science studies to understand cold fusion as undead science?

The argument of this chapter is meant to lay the groundwork for further theoretical and comparative analyses of cases like cold fusion, as well as illuminate the dimensions and benefits of a methodologically hauntological as opposed to a symmetrical perspective for science studies and the sociology of knowledge. The following discussion is meant to provide a theoretical justification for rejecting an analytical and methodological perspective that can see cold fusion only as either dead or alive.

## *Accounting for Cold Fusion*

Today, the vast majority of scientists certainly believe that the controversy over cold fusion has ended and that the phenomenon reported by Fleischmann and Pons and others was most likely due to experimental artifacts of some kind. In the eyes of many scientists, the controversy ended before it had even begun. Fleischmann and Pons were simply wrong. Yet at the same time, there are still a few hundred scientists around the world actively engaging in cold fusion research. Can we say that the controversy is over if this is the case? Historical and sociological case studies of controversy in science seem to suggest that we cannot simply look to the consensus of scientists for answers. It is always possible to find some scientist who disagrees; the question is whether anyone else is listening.[2] Buoyed by a perspective of methodological symmetry, science-studies analysts must scan the scientific horizon for signs of any continuing dispute over cold fusion. If we find nothing, then we can safely assume that the controversy has all but ended.

In 1993, Fleischmann and Pons published their heat-after-death observations in *Physics Letters A* (see chapter 5). Looking more closely at the event, we notice that the response to the article from the scientific community was generally silence (especially compared to the response to Fleischmann and Pons's work in 1989). The editors received a few letters complaining that the article had no place in a respected journal of physics. This seems to be a good sign that the controversy has ended.[3] Yet at the same time, the publication of the article was reported briefly in the media, and at least one skeptic decided to reengage with the debate (Morrison 1994).

Similarly, since 1990, anywhere from one hundred to three hundred researchers have managed to attend the International Conferences on Cold Fusion. The fact that a cold fusion article was published in a respected physics journal and that attendance is fairly stable at cold fusion conferences might give us pause in our attempts to attribute and account for closure. Perhaps research on cold fusion has become normal science and the majority of scientists are simply wrong, or perhaps most scientists are just unaware that an interest in cold fusion still exists that is not limited to one or two notoriously stubborn, "irrational" individuals.

Whatever the case, events like cold fusion conferences and the occasional publication seem sporadic and ephemeral when viewed against the background of normal scientific culture. It is not clear that the signal of scientific life is strong enough to understand cold fusion in terms of an ongoing controversy. What is more, this view creates serious difficulties for a sociological understanding of science in which scientific knowledge and truth, understood in constructivist terms, are a property of the normalized beliefs and/or practices of a majority (or even an elite). Searching the scientific horizon for controversial moments may not be as appropriate as looking at the broader space in between, the routine day-to-day activity of scientists.

An occasional journal article or even a conference may not constitute enough empirical support to resurrect the study of cold fusion. In principle, controversies may be extended indefinitely; in practice they are not. At some point we must reach a limit. There will always be a number of scientists who resist closure in controversies, but that should not prevent anyone from making a judgment about who is right or wrong. The problem is that the condition that determines this limit is a matter of how often we see or hear those dissenting scientists or others who continue to disagree. If these actors are not present, or if they are present but we cannot see or hear them, how can we account for the limit that marks closure?

## *A Convergence Theory of Knowledge*

Constructivist approaches in science studies generally, and in the sociology of scientific knowledge more specifically, can be understood in terms of what one might call a convergence theory or model of knowledge making. (There are few formal theories in science studies.) This perspective suggests that what counts as certified knowledge for societies or social groups is a product of the situated coordination of goals and interests, commitments, materials, practices, and beliefs amongst diverse kinds of actors. This is at the core of understanding the production of scientific knowledge as a fundamentally social phenomenon. However, the limits of this kind of convergence thinking become apparent when viewed in light of the case of cold fusion.

One clear statement of this convergence perspective can be found in Shapin and Schaffer's study of the constitution of early modern natural-philosophical knowledge about pneumatics: "The establishment of a set of accepted matters of fact about pneumatics required the establishment and definition of a community of experimenters who worked and shared social conventions:

that is to say, the effective solution to the problem of knowledge was predicated upon a solution to the problem of social order" (Shapin and Schaffer 1985, 282). For Shapin and Schaffer, situated social order is a condition for the certification of knowledge. The social order being postulated draws on assumptions about the convergence of actors around shared conventions of practice and notions of community, stability, and identity. The social order of communities becomes a condition of knowing (of what it is to know) in a fundamental epistemological sense.

Note that the issue is not about establishing consensual belief in a manner that reduces knowledge to a condition of collective will (i.e., what is true is whatever we agree will be true). Rather, the social condition of knowledge is the establishment of shared conventions for evaluating truth claims. Truth does not therefore reside in the procedures of transcendental justification of the kind strived for by some rationalist philosophers, but is rather a consequence of the local achievements of actors in defining the boundaries of what counts as legitimate knowledge and who counts as a legitimate knower in a given cultural context. Further, that such achievements often transcend the local conditions of their origin is a matter of their contingent extension in time and space by specific actors, and not of any inherent property of the achievements themselves.

It is for this reason that the sociology of scientific knowledge is pursued in fundamental opposition to traditions in rationalist epistemology. Scientific knowledge is generally held to be the most historically and culturally transcendent of knowledges and is viewed as being "successful" for this reason. Yet the success of science viewed sociologically is success not in terms of providing a more accurate picture of nature, but rather in terms of its situated and seemingly hegemonic extension through time and space, an extension that is facilitated by its particular forms of social organization (Latour 1993). The counter to this is that successful extension occurs only in virtue of the inherent qualities of knowledge being extended, and it is at this point that the question must turn on empirical evidence. Particularly helpful in this regard are studies of scientific controversy.

Empirical studies of scientific controversies (or controversy studies) have typically tried to advance two kinds of arguments. The first focuses on the origins of controversies with a view to highlighting the social and historical contingency inherent in the process of making scientific knowledge. The second is concerned with how stable knowledge can be produced from conditions of social and historical contingency. In sociologists' attempts to establish the conceptual foundations for the field of science studies, the first argument has typically been used as a wedge to make way for the possibility of the second.[4]

The first component of controversy studies, demonstrating radical contingency through the application of methodological symmetry, is meant to provide "empirical documentation of the interpretive flexibility of experimental results" (Collins 1983, 95). This is meant to oppose the view that the success of science rests in the inherent qualities of scientific data (or theory for that matter). Here, the notion of interpretive flexibility follows empirically from scientists' demonstrated ability to interpret experimental observations and data in different and often contradictory ways. The key point is that these differences in interpretation arise not because some actors are right while others are wrong, but because experimental data do not speak for themselves. This is to say that there is nothing about the data in themselves that can *determine* the correct interpretation.[5]

Recent debates on this point between sociologists and philosophers have demonstrated that the matter is far from settled (Franklin 1994; Collins 1994). Moreover, the participation of scientists in the debate further confuses the issue (Wolpert 1992). Despite this, however, a central guiding assumption of this book has been the inherent interpretive flexibility of experimental data and the a priori undecidability of the fact of the matter in the case of cold fusion. Of course, especially with respect to the case of cold fusion the debate rages on (see McKinney 1998; Pinch 1999), but I am sidestepping the issue for the moment in order to push the arguments stemming from controversy studies farther in the direction of understanding scientific knowledge as being constitutively social. It is time to leave the debate with traditional epistemology behind.

The fact is that the notion of interpretive flexibility has proved useful for generating symmetrical explanations of scientific disputes that allow for a full-fledged sociology of scientific knowledge rather than just a sociology of scientific error.[6] Accounting for various interpretations and the ways that they are performed in social settings provides an explanation of how and why scientific claims can be controversial without having to rely on an assumption that some actors' interpretations must be a priori irrational and/or incorrect. The assumption of interpretive flexibility has made the argument of this book possible.

Drawing on this assumption, the second element of controversy studies proposes that the correctness of a given interpretation, its truth or validity, should be understood as a situated social or cultural achievement. After the first, essentially deconstructive move, controversy studies are meant to focus on "the ways that the limitless debates made possible by the unlimited interpretative flexibility of data are closed down" (Collins 1983a, 95). This attention to the mechanisms of closure forms the basis of the constructivist epistemology in the sociology of scientific knowledge. Truth in scientific controversies is neither absent nor unattainable; it is constructed through the social and material mediations or negotiations of various actors. Truth or correct interpretation is synonymous with the closure of controversies; closure results in statements and objects that scientists come to deploy as true or real.

As Shapin and Schaffer (1985) and Shapin (1994) argue, the adjudication of the correct and incorrect interpretation of data is predicated on the establishment of some kind of social-material stability, understood as a "form of life" that makes the process of collective adjudication possible. Given this perspective, controversies are usually treated by constructivist analysts as rare instances of disruption, disorder, and disorganization. These may be interpreted in terms of major historical ruptures or revolutions (Kuhn 1970), or in a more microsociological framework, controversies may be treated as more or less local problems of social coordination that alter social networks and practices over time (Barnes 1982; Callon 1980).

## *Mapping the Dynamics of Controversies*

At the core of conceptions of closure in controversy studies is the idea of convergence. Dissension gives way to agreement. Conflict and disorder give way to stable communities and routine practices. This can be represented visually.

Figure 7.1 is a reproduction of Martin Rudwick's schematic representation of the pattern of the controversy amongst nineteenth century geologists over the existence of the Devonian period in geological history (Rudwick 1985, 412–413). Rudwick's scheme is too complex for extended discussion here, but I want to draw attention to the general funnel shape of the pattern. In Rudwick's view, the controversy was resolved as the divergent positions of key actors in the 1830s converged on a general consensus around 1839. This is represented in Rudwick's schema by the relative separation of the lines or "interpretive trajectories" of the various actors over time. The convergence of the lines represents the putative end of the Devonian controversy and also the end of Rudwick's story.

The characterization of the consensus position is not as important here as the assumptions present in the schematization of the movement from controversy to closure. As Rudwick writes, "As a matter of history, the controversy ended as soon as a set of less than ten leading geologists in the early 1840's converged in a collective judgement that the problem had received a satisfactory solution" (Rudwick 1985, 428). Sociologists and historians may differ in their opinions as to whether closure can be attributed to so few actors, but almost none would question the sense that some kind of collective stability was achieved.[7] Rudwick is concerned to emphasize the convergence of the interpretive judgments of a few key individual actors, but any number and any kind of actors can be modeled in the same terms.

Even with his de-emphasis of human agency, Latour characterizes closure as the production of black boxes in similar terms of convergence: "The simplest means of transforming the juxtaposed set of allies into a whole that acts as one is to tie the assembled forces to one another, that is, to build a machine. A machine, as its name implies, is first of all, a machination, a stratagem, a kind of cunning, where borrowed forces keep one another in check so that none can fly apart from the group. . . . The assembly of disorderly and unreliable allies is thus slowly turned into something that closely resembles an organized whole. When such a cohesion is obtained we at last have a black box" (Latour 1987, 129). The dynamic portrayed here is a movement from an assembly of "disorderly allies" (humans and nonhumans) to a coherent organized whole that works like a machine. Latour's machine is not different in principle from Rudwick's "collective judgment." Both give an account of the constitution of facts in terms of convergence.

A more simplified and generalized version of Rudwick's diagram is shown in figure 7.2. The beginnings of controversies are typified by a maximum of interpretive flexibility, social instability, and cognitive divergence. As a controversy progresses, actors convince, influence, mediate, and force each other to alter their positions, drawing on a variety of social and material resources. Over time, the various actors fall into line or converge around a single representation, theory, practice, or design, and the controversy comes to an end. The period following this convergence or closure is a period of routinization, in which all actors accept the facts or deploy the objects unproblematically. At the point of routinization convergent networks of actors may be described as black boxes and as such become naturalized facets of the social and material world.

Support for this more generalized model of the development of controversies can be found in Latour, Mauguin, and Teil (1992), who present another form of mapping controversies that they call "socio-technical graphs."[8] The specifics of their method of mapping differ from mine, but the conclusion in terms of implications for analysis is the same: "If several accounts converge, and if the actants they mobilize have a high degree of coherence, then the degree of predictability of the project increases. At the limit it might even be possible to predict the next move. If on the contrary, there is a high degree of dispersion among accounts, and if actants they enrol have no stable

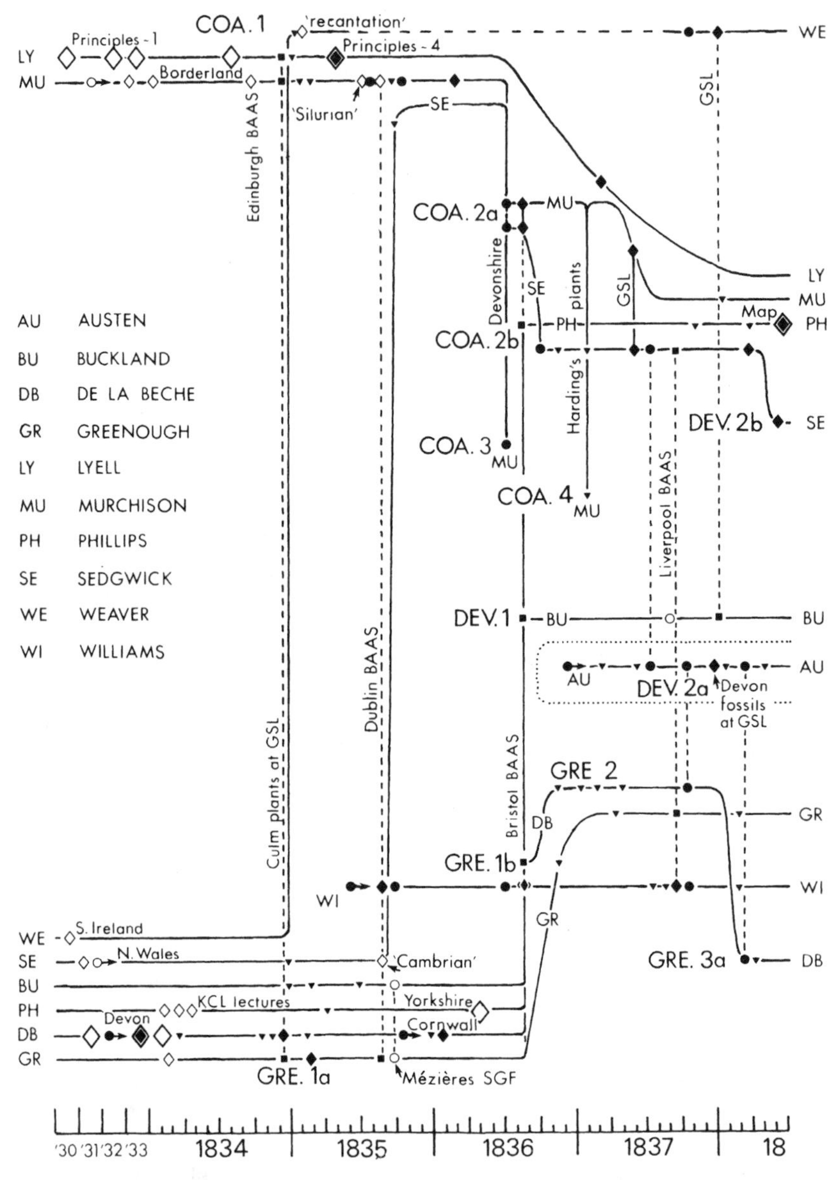


*Figure 7.1* Convergence in the Great Devonian Controversy. *From* The Great Devonian Controversy: The Shaping of Knowledge among Gentlemanly Specialists, *by Martin Rudwick (Chicago: University of Chicago Press, 1985), 412–413.*

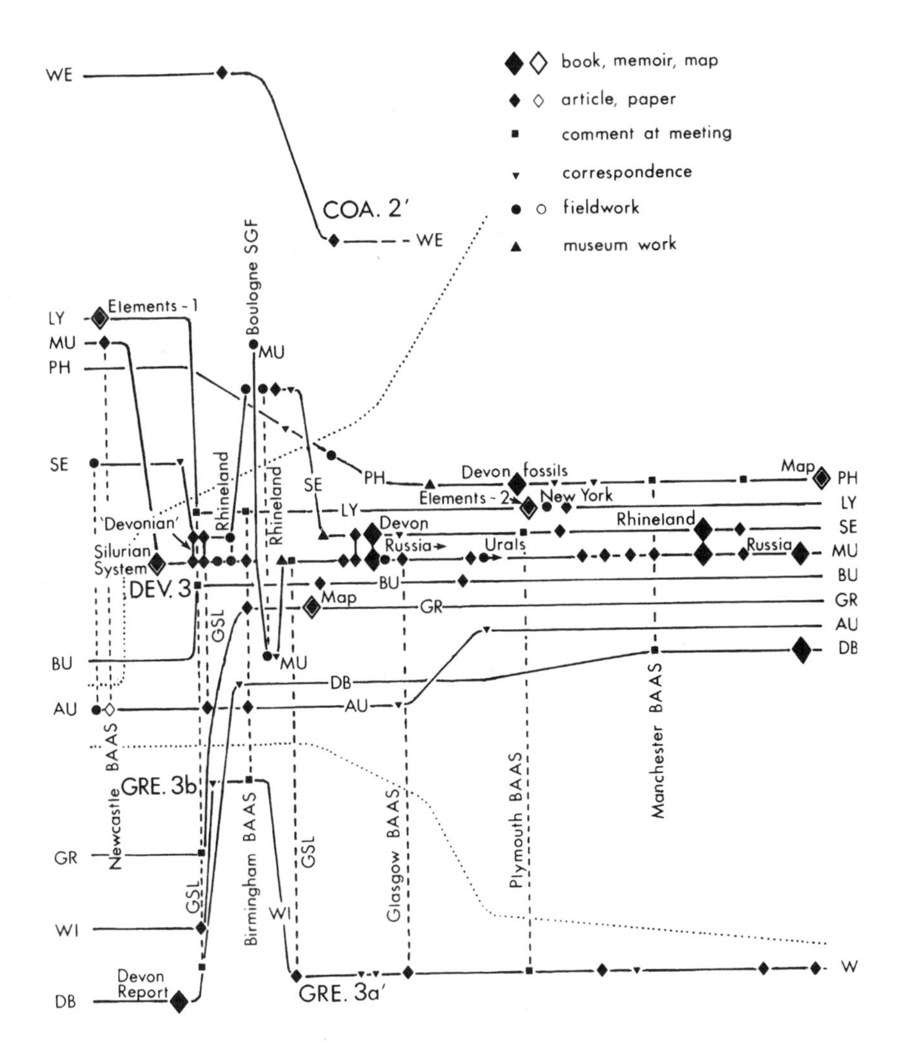

book, memoir, map
article, paper
comment at meeting
correspondence
fieldwork
museum work
WE
COA. 2′
WE
LY
Elements - 1
MU
PH
SE
Boulogne SGF
MU
Rhineland
Rhineland
SE
PH
Devon
fossils
Map
PH
LY
Elements - 2
New York
LY
'Devonian'
Devon
Rhineland
SE
Silurian System
Russia
Urals
Russia
MU
DEV. 3
BU
BU
GSL
Map
GR
GR
AU
BU
MU
DB
DB
AU
AU
Newcastle BAAS
GRE. 3b
Birmingham BAAS
GSL
Glasgow BAAS
Plymouth BAAS
Manchester BAAS
GR
GSL
WI
WI
Devon Report
DB
GRE. 3a′
W


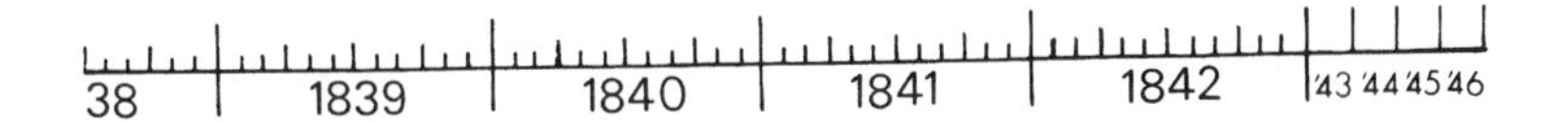

38
1839
1840
1841
1842
'43 '44 '45 '46

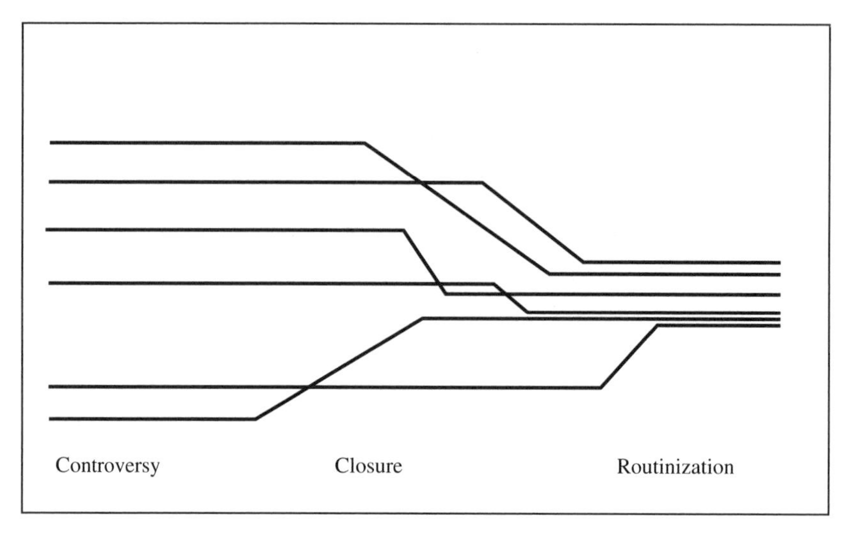


*Figure 7.2* A Simplified Convergence Model

definition, the interpretive flexibility will be so great that no prediction is possible" (Latour, Mauguin, and Teil 1992, 50).

We can understand convergence in strictly Latourian terms as the process whereby actor networks gain allies, thereby increasing their capacity to resist dissension and compel assent. The greater the convergence, the more difficult it is to maintain a controversy: "The practical limit is reached when the average dissenter is no longer faced with the author's opinion but with what thousands and thousands of people have thought and asserted. Controversies have an end after all. The end is not a natural one, but a carefully crafted one like those of plays or movies" (Latour 1987, 59). Rudwick's diagram combined with Latour's actor-network models nicely captures our sense of the stability and universality of scientific knowledge while illuminating the ways that knowledge may be destabilized in different contexts where the degree of convergence amongst actors is not as great (i.e., where networks are weaker or less far reaching).

Even where convergence is not total (i.e., there are still actors in a controversy who continue to dissent), the convergence model helps makes sense of the position of the dissenter or dissenters without needing to take a position of epistemological relativism that denies the contingent truth of the matter. This notion of convergence is asymptotic. Actors do not converge on the truth; the truth, as such, is a product of their convergence. This convergence is never complete or absolute, so reality for actor-network theorists is understood abstractly in terms of a gradient of force that is able to withstand, eliminate, or translate dissension.

## Convergence and the Politics of Addition

The dynamics of controversies implied by the convergence model map a function of "many to one" where multiple, dispersed actors mediate themselves into single, convergent stable black boxes. One thing to notice about controversy studies is that the movement from dispersion to convergence is primarily a process of combination or addition. Actors, interests, and objects combine or add to one another to produce stronger networks of associations. In the original group of SSK-inspired controversy studies, the notion of social agreement and alignment of interests tended to invoke the mechanics of addition (often building to a concept of consensus), and in later actor-network studies, the movements of actors (both human and nonhuman) are often described in terms of the enrollment or enlistment of actors (Callon 1986). Convergence, then, is an additive property.

If convergence is a function of addition, an issue arises when we must decide which actors to count in our calculations. Should we map the trajectories of all possible actors in a controversy? Even using a supercomputer and an army of graduate students to model such a complex set of dynamics seems hopeless, but it is also pointless. Simply put, not all actors figure equally in the production of scientific knowledge. Moreover, convergence may be better achieved by subtracting rather than adding actors to the network. Perhaps we need to think more carefully about specific kinds of translations that occur in the process of ending controversies. Some actors are added to the network, some actors are subtracted, and others are probably not translated at all.

Note that this implies a subtle reinterpretation of the notion of translation. For Callon (1986), for instance, translation is the process whereby resistant actors are turned into allies. I am suggesting that translation can also be a process whereby resistant actors are turned into enemies. Actors subjected to this kind of translation may be ignored, marginalized, and pathologized, and they may even be subjected to mechanisms of suppression or repression (Martin 1992). These are aspects of the production of stable convergence that seem to have been neglected in actor-network studies. This idea is meant to direct our attention to the dynamics of exclusion, which form part of the process of ending controversies and stabilizing knowledge.

Assume for the moment that for every actor that is enrolled or included in the convergent network, there is another actor that is excluded or disenrolled.

I would call this a translation of subtraction, and the affected actors are the forgotten and/or pathologized things and people in our science worlds; they are our "others."

It seems reasonable to suggest, however, that in a given controversy not all actors converge in the same network. Some, for a variety of reasons, will remain resistant to the pressure of inclusion. Expanding on the agonistic metaphors of actor-network theory, we could call those who resist enlistment "conscientious objectors." Perhaps there are also actors that seem to go along and then leave the network or switch sides; such actors might be called "defectors" or "traitors." Finally, there are those actors who simply do not qualify for enlistment; perhaps these actors have the equivalent of "flat feet" as far as the network is concerned. What happens to those who end up outside convergent networks? The suggestion is that such actors constitute the "losers" in the case studies of scientific controversy, and that these losers exist alone or in weak associations. The "winners" are convergent, and the "losers" are seen as stubbornly resistant, divergent, and dispersed, a condition in which it becomes literally impossible for dissenters to "get their act together." The losers lost, after all, because they failed to marshal the strength and coherence to dissociate the networks of their opponents.[9] This may sound fine, but how do we recognize it? How do we recognize convergence?

## *Measuring Closure*

In many ways the dynamics of the cold fusion controversy do follow this modified convergence model, as demonstrated by portions of the story I told in previous chapters. In 1989 the controversy was characterized by rationally interminable arguments over what experiments if any could count as successful replications of cold fusion. The controversy ended in favor of the critics, whose solidarity in understanding cold fusion in terms of conventional nuclear and plasma physics established stable criteria by which would-be replicators might judge their data. Interpretive charitability for CF claims diminished, and without strong evidence of nuclear ash along with excess heat, there was no real justification to continue. By the end of 1990, as more and more scientists started losing interest in cold fusion, skeptics and boundary workers were able to invoke the absence of experimental evidence for CF as evidence for CF's absence. Negative results were interpreted as being positive evidence that cold fusion was not real. Researchers working at MIT's Plasma Fusion Center, for instance, stopped doing their CF experiments and held a mock wake for cold fusion (complete with black armbands). A comment made by John Maddox (a former editor of *Nature*) in a BBC documentary on cold fusion in 1990 reflects the shared perception that the controversy had ended: "I think it will turn out after two or three years more investigation that this is just spurious and unconnected with anything that you could call nuclear fusion—thermonuclear fusion. I think that, broadly speaking, it's dead, and it will remain dead for a long, long time."[10]

Of course an even better indicator of closure was not the prevalence of opinions like those of Maddox or the MIT researchers but rather the decrease in public challenges to these opinions. The cold fusion controversy was not over simply because a number of important scientists said it was. Although such comments may play an important role, it makes sense that controversies do not end by fiat. Instead, we can view the appearance of these comments at the end of 1989, in 1990, and later as an indicator that the controversy is over (or all but over). There are more material indicators of closure as well. One of the most accessible can be found in the quantitative and semiquantitative analysis of trends in the scientific literature. As Lewenstein (1995, 420) has shown, the number of media and technical publications on cold fusion sharply decreased by the end of 1990. An analysis of the literature in terms of the submission dates on the articles further indicates that most of the research on cold fusion had been done in 1989, and far less new work was being subsequently submitted (and accepted) for publication.

Perhaps an even more forceful indicator of closure is the decline in resources for doing cold fusion experiments. In mid-July 1989, the panel of experts appointed by the U.S. Department of Energy (DOE) to evaluate cold fusion released its interim report recommending against any special federal funding for research. The panel had visited several laboratories in the United States and had decided that insufficient evidence had been accumulated to justify further expenditures by the federal government. The final report appeared in November, and soon thereafter the various national laboratories with cold fusion research programs were directed to shut them down. Not long after this, other sources of funding dried up. The $5 million allocated for the National Cold Fusion Institute in Salt Lake City ran out in 1991, and the institute was closed as the anticipated partnerships with industry did not appear. In later years, standard applications for funding from the DOE were also being turned down. Other sources of revenue for research were cut off as applications for patents related to cold fusion were rejected by the U.S. Patent Office. The University of Utah's applications for patents on cold fusion have never been accepted, and many researchers who claim to have positive results but have no way to develop them are also unable to get patents on their work.

It also became increasingly difficult for researchers to make use of the resources they did have access to. University deans and department heads discouraged further cold fusion work, and scientists felt pressure to return to more "promising" research. In one case, the tenure of a professor at MIT was threatened (Pool 1989). More importantly, the chief source of scientific labor, graduate students, disappeared as cold fusion became an increasingly problematic dissertation topic. As they were deprived of resources and labor, it became more and more difficult for CF researchers to improve or even maintain their situation.

More indicators of closure can be found not in the destruction or erosion of support for cold fusion research, but in the reconstruction of cold fusion as a case of pathological science. In February 1990, Douglas Morrison published the first participant "history" of cold fusion in the popular magazine *Physics World*. The article, entitled "The Rise and Decline of Cold Fusion," makes use of Irving Langmuir's famous lecture on pathological science to describe scientists' continuing belief in the existence of cold fusion as an instance of bad scientific judgment. For Morrison and a growing number of scientists, the question was no longer whether cold fusion was real, but why it was that any scientist could have believed that it was real in the first place.

The hypothesis of pathological science became a credible explanation for why some researchers still claimed to be getting positive results. These researchers were in effect behaving pathologically, and many of them became marked or stigmatized as pseudoscientists, cranks, or even criminals. In what might be a landmark case, an Italian court ruled in favor of the newspaper *La Repubblica* in a libel suit launched by Fleischmann and Pons and Italian cold fusion researchers. The judge ruled that given the abnormal behavior of scientists in the cold fusion affair, the newspaper was justified in accusing Fleischmann and Pons of fraud. This judgment is an excellent example of what has come to count as a normal explanation in the case of cold fusion.

All these indicators serve as evidence for the attribution of closure in the case of cold fusion. The implication here is that any further study should be concerned not with rare controversial events but rather with what scientific life is like now that cold fusion is dead, buried as it were in a black box of pathological science. One chemist I interviewed in 1996 told me, "Before your call, I don't think anybody has actually even mentioned cold fusion to me in some three or four years. It's a nonissue here."[11] This sentiment is echoed by the majority of the world's scientists, who, if they consider cold fusion at all, consider it as a historical oddity in the normal progress of science. We are left with the impression that as far as normal science is concerned, cold fusion is of interest to no one except crackpots, pseudoscientists, con artists, and a few sociologists of science.

## *Considering Cultural Relativism*

This brings us back again to our dilemma. Assuming that we agree to reject the hypothesis of pathological science as being epistemologically problematic (because Langmuir's criteria do not actually apply in the cold fusion case) and sociologically asymmetrical (because one side is assumed to be necessarily irrational while the other is not), the continued presence of cold fusion researchers remains bothersome, especially since it is unlikely that they will disappear anytime soon. Because of this, at least in the case of cold fusion the convergence model appears flawed. In the course of this book I have attempted to move beyond the limits of analysis imposed by methodological symmetry, which states, that the outcomes of controversies could be otherwise. Instead I have tried to show that in the case of cold fusion, the outcome *is* otherwise. While we now, as a result of the controversy, live in a world where cold fusion does not exist, research on cold fusion continues, and knowledge is still generated about the phenomenon. Crucially, CF researchers are not just a weak association of individual dissenters but rather they constitute a coherent (possibly convergent) network that survives in the interstices of what we recognize as normal science worlds. It is in this sense that I want to argue that CF research can be characterized as undead science.

One viable solution to the mystery of the afterlife of cold fusion, which may answer the question of whether we see controversy or closure, is a metaphysical commitment to cultural relativism. The notion that other cultures can rationally hold alternative or even contradictory beliefs about the world is no longer a problematic idea in the social sciences. A controversy that is closed for us (in "our" culture) might conceivably be open or even closed differently for others (in another culture). This might, for instance, explain the interest in cold fusion in Japan. A culturally relative argument prompted by methodological symmetry might suggest that since Japanese culture (or at least the organization of Japanese science) is different from North American culture, the outcome of the cold fusion controversy could be, and perhaps is, different there. In this sense the Japanese become like a culturally distant Azande (Evans-Pritchard 1937), or even a closer-by millenarian cult (Festinger, Riecken, and Schachter 1956).

The same could also be said of cold fusion researchers in the United States: their differing beliefs might be taken as a sign that they belong to another culture that is not a part of science, or is not scientific. What we would have then is not the death of cold fusion, but closure through the bifurcation of scientific and pseudoscientific subcultures. Thus cold fusion is dead for *us* as members of properly scientific worlds but not for *them* as members of pseudoscientific or even just differently scientific worlds. This multiple-worlds thesis is certainly a plausible alternative to the technoscientific afterlife, but such a thesis may be less sensitive to the interactions and inequalities that characterize the relationships of social worlds in the context of increasingly globalized and public scientific practice.

## *The Importance of Asymmetry*

Rather than pursue the relativistic argument, perhaps we should consider a more realist response to the question of whether cold fusion is alive. We no longer hear about cold fusion because cold fusion is not real. If there are scientists who still believe in the phenomenon, they are wrong, and we can invoke explanations of their behavior in terms of irrationality, pathological science, fraud, or mass delusion if we like. Clearly, this asymmetrical response is typical of opponents of cold fusion and more than a few commentators, including sociologists of science, but it is also an attitude prevalent in Western European and North American culture.

Cold fusion is in a crucial constructivist sense very solidly not real. It may not be quite as unreal as the flatness of the Earth, but it is unreal enough that it won't raise eyebrows if it becomes (as it has) an example of how not to do research in books on ethics in science. We must take realist asymmetry seriously because, as members of a society that tends to trust the proclamations of recognized experts, we must be asymmetrical also. It is, as we know, an important part of going to the doctor, flying on a plane, and indeed maintaining social order generally. Not only is it not wise for science-studies analysts to invest their life savings in palladium futures, to do so would be to ignore a crucial part of the story.

As I explained in chapter 1, there is nothing novel about this point with respect to the sociology of scientific knowledge; I am simply drawing attention to the social and material force of practices and beliefs around what comes to count as real (and not-real) in the resolution of controversies. In the asymmetrical moment we can "feel" the tangible social and material consequences of not believing the diagnoses of doctors or, alternatively, believing the claims of cold fusion researchers. In the latter case, if you are a scientist, your professional life may suffer. If you are an investor, you are bound to lose your money. If you are a science-studies analyst, you may be hoaxed by Alan Sokal (1996).

This brings us back to the methodological double take I described in chapter 1. My question is whether we can methodologically accept a symmetrical story of closure and employ it with the conviction of an asymmetrist. Note that this is different from the standard methodological route of alternation, which requires us to be only as symmetrical as the scientists we study. The question is whether, as analysts, we can be both symmetrical and asymmetrical at the same time. Why not? Symmetrical historiography and Whig historiography have at least that much in common. No matter who tells the story or how it is told, Pasteur won against Pouchet in the debate over spontaneous generation (Conant 1951; Farley and Geison 1974; Latour 1989). Boyle won against Hobbes in the controversy over the existence of the vacuum (Shapin and Schaffer 1985). Blondlot lost in his bid to prove the existence of N-rays (Nye 1980), and Fleischmann and Pons lost in their battle to keep cold fusion alive. The current configuration of Western knowledge and society is a testament to these outcomes, and there is little reason for sociologists of science to reject them.

One wonders, however, observing the ongoing machinations of cold fusion researchers at the same time as they are being ostracized, whether it is possible that other "losers" have similar experiences. Whatever happened to Pouchet and Blondlot and the many others who engaged and lost in the supposed struggle to produce truth? It is interesting to speculate on the possibility of a vast underground network of scientific research. Perhaps the case of cold fusion is just the tip of the iceberg? I want to guard against the pluralist and symmetrical theorizing of cultural relativism, which could dispense with this question by observing that the "losers" of controversies, to the extent that they survive, simply go on doing their own thing. Cold fusion researchers are not Azande, and they are not cultists, nor are they freaks. My point is that they are not so different that they can be made not to count in stories about the construction of *our* scientific knowledge.[12] Cold fusion research in this sense is not a parallel world of rejected science but is intimately intertwined with the institutions and processes that produce the scientific knowledge and technological objects we are used to.

One of the central arguments of this book is that the closures of controversies have important social and epistemological effects on the organization of scientific knowledge and practice. This is a relational but not a relativistic point. That is, in the case of cold fusion, closure as a process produces reliable knowledge about the world (in this case the knowledge that cold fusion does not exist) in relation to a collectively sustained alternative belief embedded in the material practice of science. CF researchers do not see the world differently, nor do they adhere to an alternative incommensurable paradigm. Their beliefs and practices are held in constant relation to the organization of belief and practice in mainstream science, and the social world of cold fusion research is constructed in relation to (not relative to) this organization.

As has been previously observed with parapsychologists (Collins and Pinch 1982), most cold fusion researchers consider themselves to be scientists (indeed many consider themselves to be better scientists than their critics). Moreover, many cold fusion researchers are in fact accredited professional scientists and engineers at prestigious universities. We need to look closely at who cold fusion researchers are and what they do before we can say that wherever we see research on cold fusion, we must see something that is not a part of what we should understand to be normal scientific practice and culture.

As I tried to show in chapter 5, research on cold fusion after 1990 proceeds much like more legitimate scientific research. While the number of researchers has dwindled since the initial replication effort of 1989, those who remain make up a core group committed to the investigation of the phenomenon and the development of the field. Indeed, when compared to some specialized areas in other scientific fields that may involve no more than a handful of research groups, the number of scientists currently working on cold fusion seems quite ordinary.

As I have mentioned, a relatively large research effort is continuing in Japan in both public- and private-sector laboratories. Experimental and theoretical work of different kinds is also being conducted at several labs in Italy, France, China, Russia, and India. In the United States, individuals and small groups are active at places like Los Alamos National Laboratory, Texas A&M, MIT, the University of Illinois, and Stanford Research Institute. While many of these scientists have been continuing with work they started in 1989, new individuals have joined the effort, and new experiments have been performed.

Despite the decline in submission and publication of papers on cold fusion in many reputable journals, work on cold fusion does get published. The most prominent forum has been, ironically, the *Journal of Fusion Technology*, which until recently featured a regular section devoted to cold fusion research alongside engineering articles related to conventional hot fusion. To date, one bibliography of technical literature has over a thousand references, with a little less than half of the articles being published in 1991 or later (Britz, n.d.).

Research on cold fusion has not been limited to reproducing the original experiments of Fleischmann and Pons. Prompted mostly by the secrecy of those two researchers, many other researchers have developed their own ways of producing the cold fusion effects; these include new kinds of apparatus and procedures for preparing the electrochemical cells, as well as methods of loading deuterium in palladium that do not necessarily involve electrochemistry. Some of these researchers continue to measure quantities of excess heat and nuclear particles with increasing accuracy and reliability, while others struggle to reproduce a stable effect. In addition, a number of new effects related to Fleischmann and Pons's original claims have been observed. The most startling of these are reports of the measurement of excess heat and nuclear particles (mostly tritium) using light-water-based electrolytes with nickel cathodes, as opposed to heavy water and palladium. Other researchers have reported the production of helium, X-rays, and large bursts of neutrons, and most recently a number of researchers have started pursuing measurements of nuclear transmutations in their experiments.

Despite the proliferation of experimental techniques, there is no consensus amongst cold fusion researchers about how to explain the phenomena, or even about the methods and procedures one must use to produce the effects reliably. They agree on the legitimacy of investigations into cold fusion phenomena and the existence of some kind of anomalous observations, but they continue to disagree about what the phenomena amount to. Certainly post-closure cold fusion cannot be described in terms of a stable research program or paradigm. The organization of research seems preparadigmatic, more like a fledgling scientific field.

This would be the case perhaps, if it were not also for the fact that while cold fusion research does go on through various scientific institutions, a great deal of work is performed outside normal disciplinary spaces, in the garages and basements of people's homes, in privately circulated newsletters, and in e-mail newsgroups and Internet web pages. A closer look at who attends the annual cold fusion conferences reveals a mix of professional scientists and retired, semiretired, and amateur scientists, engineers, and technicians, as well as a number of entrepreneurs, inventors, and interested lay people. Topics of discussion at the conferences range from technical issues in the measurement and analysis of excess heat to the relationship between cold fusion and even more marginalized practices like alchemy or the design of perpetual-motion devices.

The result has been the development of an extremely heterogeneous research world that appears to be normally scientific and utterly unscientific at the same time. Marginalized scientists working on experiments started in 1989 find themselves communicating and collaborating with researchers working in even more esoteric and illegitimate areas. At one conference I attended, a couple of men with no significant scientific or technical training arrived with a metal device that they claimed had the capacity to produce cold-fusion-like transmutation effects. Their device had actually been tested by two scientists, who had observed some anomalous results, but as these researchers discussed their data, they claimed to have received the instructions for building their device from God. In normal science worlds it would not necessarily be unusual to receive inspiration from on high (Noble 1992), but it would seldom be accepted as a legitimate form of the experimental protocol.

## *The Relevance of Ghost Stories*

This observation is borne out in the literary genre of traditional ghost stories (and in recent films like *The Sixth Sense* [1999] and *The Others* [2001]. Ghosts are strange and uncanny, but they are not monstrous alien others. They do not come from elsewhere; rather, they belong to us and walk among us as part of our culture. They are often old friends and relatives that occupy our houses. What makes a ghost story scary is more the effects of recognition than the fear of the unknown. This is an observation advanced by Freud in his famous essay "The 'Uncanny'" (*unheimlich*), where the phenomenon is described as "that class of the terrifying which leads back to something long known to us, once very familiar" (Freud 1958, 124). In Dickens's *A Christmas Carol*, Ebenezer Scrooge discovers this when his dead friend Marley first visits him. Like Marley, CF researchers represent the ghosts of familiar associations thought to be long gone and yet remaining. Indeed it is the perception or imagination of their unfamiliar presence in familiar contexts that for Freud becomes a source of horror or anxiety.

It is the closeness or familiarity of CF research to normal science and scientists that prompts an understanding of post-closure cold fusion in terms of undead rather than rejected or marginalized science. The dynamics of rejection and marginalization suggest a distancing or removal of subjects from the legitimate arenas of science. Rejected sciences become increasingly unrecognizable to normal scientists as science and therefore are less troublesome and less of a cause for concern. While closure and pathologization in the case of cold fusion have indeed produced these effects, CF researchers (at least core-group members) continue to remain a part of normal science worlds. This is the sense of CF research as being both inside and outside science at the same time, a sense that I tried to convey in chapters 5 and 6. This social-epistemological position that CF researchers collectively occupy is what I am defining as undead science. As I intend for the concept to have some comparative utility, let us consider once more the properties of ghostliness as they apply to the case of post-closure cold fusion.

### A RELATION TO DEATH

Throughout this book I have made the antirelativistic point that post-closure cold fusion research is science but not normal science. The various undead creatures of literary and pop-cultural imagination, particularly ghosts (but also vampires and spirits), may act like normal human beings and even pass for normal human beings from time to time, but their lives are forever marked by the experience of their death. In many stories, the souls of the departed remain as ghosts to avenge their untimely death or to serve as a warning to the living, or because they have not recognized that they have died. Whatever the case, the life of a ghost is always circumscribed by the events of its death. It is after all the moment of death that produces the ghost as a post-closure entity in the first place.

I have tried to argue that this is how we should understand post-closure cold fusion research. Both the identity of researchers and their experimental practice have been circumscribed by the events that led to the closure of the controversy in 1990. CF researchers live and work in the aftermath of that closure and with the memory of the controversy. The swiftness with which closure occurred in this case acts as a kind of traumatic death that reverberates in the activities of researchers after 1990. It is a source of endless frustration for some. No matter what kinds of new experiments and data CF researchers present, their critics (if even listening) always drag them back to debates from 1989. There seems to be no escape. As a consequence, CF researchers recognize that they lost the controversy in 1989, while refusing to believe that cold fusion is dead. Their collective identity as a group as well as their experimental practice is organized to a large degree around reclaiming scientific legitimacy by constantly revisiting the criticisms of 1989–1990. In part they have no choice, since there are few extant criticisms of work after 1990 that they can address.

## INVISIBILITY

Ghosts are invisible, or perhaps more importantly they are visible only to those who have the "gift" for seeing them. This invisibility comes not from being transparent to light so much as from the fact that ghosts tend to move in places living human beings do not normally look. Ghosts, like other undead creatures, travel at night. They appear in out-of-the-way places like graveyards and attics, and on deserted streets. While they are occasionally detected and detectable in more familiar settings (in séances and hauntings), for the most part they pass unnoticed even when they sit beside us at the dinner table. In any case, the general impression we are left with in both parascientific and literary studies of the undead is that the few that are detected are merely a small portion of the world's ghost population.

As I have argued, for the majority of scientists cold fusion research is for the most part invisible. Most scientists believe the controversy to be long over, and so they are not looking to see if CF research is occurring in the first place.

(This is why so many scientists I have talked to have been surprised to learn that there are so many scientists still working on the topic.) Core-group CF researchers tend to operate quietly and in out-of-the-way places. Perhaps they are less visible because they work under the cover of more respectable labels like "anomalous effects in deuterated metals," but it is also because they often work on their own time (often in their own homes) outside the routine activities and relationships that form the bulk of their working day.

For some CF researchers, their invisibility stems from working outside the boundaries of their normal career path. Many CF researchers, like John Bockris and Martin Fleischmann, are retirees; freed from the pressures of their careers, they are more able to engage in speculative research, but by the same token the work that they do becomes less visible to their still employed colleagues. Finally, even those scientists who strive to make their CF research more visible often have difficulty because the resources for enabling visibility (peer-reviewed publications, patents, grants, media exposure, etc.) are not routinely available to them. All of these features of the post-closure world lend themselves to describing CF researchers in terms of an underground or submerged community, but the invisible work of CF research arises in this case from the memory of the end of the controversy. This memory produces a situation in which most scientists have stopped looking for cold fusion (as they perceive it to be dead), while at the same time CF researchers' efforts to make themselves visible have been curtailed (due to lack of resources or fear of being stigmatized).

### MOBILITY

Ghosts can walk through walls. The conventional social and material boundaries that structure the activities of living human beings do not apply to them. The behavior of living human beings is structured both by the materiality of the world in which they live (we cannot walk through walls) and by the norms and conventions of social interaction. In virtue of being dead, ghosts are less bound by these restrictions. They are ephemeral, able to move fluidly between different kinds of social and material spaces. It is important to note, however, that in many literary accounts of the movement of ghosts, they are not able to go anywhere or do anything. They have greater mobility than living human beings, but they are still tied to the corporeal world. In some stories ghostly mobility can be highly circumscribed: poltergeists are bound to a specific material space such as a house or a piece of land.

Similarly, CF researchers have a mobility that most scientists do not have, while they nevertheless remained tied to the materiality of concrete experimental practice. As I noted in chapter 6, the lack of normal institutional resources for doing CF research has led core-group scientists to take advantage of the different kinds of resources of the cold fusioneers. Normally scientists would not consider publishing their research in a popular magazine like *Infinite Energy*, but that magazine has become a forum for the communication of technical data (amongst other things) that CF researchers find valuable. While they struggle to publish in mainstream peer-reviewed journals, they also make use of less legitimate media such as newsletters, magazines, and Internet websites.

CF researchers, especially core-group members, are also mobile in their ability to shift between a mainstream scientific identity and their identity as CF researchers. For many scientists and engineers working in normal scientific institutions, CF research is a part-time activity that seldom explicitly intrudes on their more routine activities. For these scientists, often working alone in the context of their institutions, cold fusion remains a curiosity, a hobby, or an interesting problem worth the investment of an amount of personal time and resources. But once these scientists enter conversation with other researchers or attend a cold fusion conference, they can become more defensive about their research and express frustration at the ignorance of their colleagues. This shift in identity represents a kind of ghostly mobility.

Finally, it is important to point out that since CF research was for a brief time a form of legitimate science, so it remains tied or bound to the norms and expectations of legitimate science. Like a spirit that is bound to the place where it died, CF researchers are tied to the normal expectations of science. They continue to perform experiments, as any legitimate scientist would do, and the goals of replicability, error correction, and explanation continue to be foremost in any experimental effort they undertake. In this way, ghostly mobility does not imply ad hoc relativism. CF researchers are not free to believe anything they like, just as ghosts are not free to go anywhere they like.

## HAUNTING

If ghosts were totally invisible and mobile, there would never be any ghost stories. Ghost stories are by and large accounts of encounters between living human beings and undead spirits. Such encounters, when they occur repeatedly, are often referred to as hauntings. From the point of view of the living, haunting is what ghosts do. In the stories there are as many forms of haunting as there are ghosts. The ghost of Marley, accompanied by the sound of rattling chains, visits Ebenezer Scrooge to warn him of the impending evening with the ghosts of Christmas past. In Shakespeare's play, Banquo returns from the dead and simply floats in the room to give Macbeth a fright. In a more contemporary vein, Steven Spielberg's *Poltergeist* films from the 1980s feature a series of powerful spirits that can rip up houses and possess little children.

Whatever the precise form, haunting is the means by which ghosts make their presence known or felt in the worlds of the living.

It is worth noting that hauntings typically generate two kinds of effects; the first is uncanny material effects. These are most common in haunted houses; scrapings, rattlings, loud bangs, moans, and cries are the standards here. In addition, there is moving furniture, the opening and closing of doors, and even knives flying through the air. As a point of interest, the most extreme form of material effect is probably spirit possession, but this explicit merging of the ghostly and the corporeal is best left for a different kind of study. The second effect of haunting is more directly social-psychological. Through either the direct presence of a ghost or its material effects, hauntings produce feelings of anxiety, fear, and horror. This is why ghost stories are meant to be scary, or if not scary then at least troubling or unsettling.

The argument of this book has been that since the closure of the controversy in 1990, the ghosts of cold fusion continue to haunt the house of science. Moreover, the activities of these ghosts produce palpable material effects as well as anxieties and fears amongst the scientists with whom they interact. In chapters 5 and 6 I enumerated some of these material effects. Fleischmann and Pons's heat-after-death paper appearing in the journal *Physics Letters A*, Miley's cold fusion section in the *Journal of Fusion Technology*, Patterson's patent, demonstration device, and appearance on the *Nightline* news program, and the Japanese New Hydrogen Energy project are all examples of the scrapings and rattlings of CF researchers whose work has seen the light of day. These events represent evidence of a more extensive presence; such hauntings are epiphenomena or apparitions of the more invisible and mobile work that goes on.

For the most part, these apparitions of cold fusion are not noticed by the majority of scientists. A few of the mainstream readers of the *Journal of Fusion Technology* may have been annoyed or even scandalized by the editor's insistence on including cold fusion articles, but most readers just ignored the cold fusion section, concentrating instead on papers more relevant to their own research. While some scientists I spoke with expressed an awareness of continued cold fusion research—through their exposure to articles in the *Journal of Fusion Technology*, for instance—many others seemed nonplussed to find the journal support claims that they believed were spurious. Yet in the case of the journal and in other hauntings, the apparitions of cold fusion produce an anxiety that prompts some scientists to view CF research as a threat or an abomination to science. In these instances, there is often an attempt to exorcise the ghost, such as the attempt to remove the distinguished-professor status of John Bockris at Texas A&M University, or the attempts by Robert Park and others to shut down cold fusion sessions at American Physical Society events. Ongoing pathologization of the kind discussed in chapter 4 becomes a consequence of the anxiety produced by the haunting of the ghosts of cold fusion.

## *Haunting Science Studies*

What may this mean,
That thou, dead corpse, again in complete steel
Revisit'st thus the glimpses of the moon,
Making night hideous; and we fools of nature
So horridly to shake our disposition
With thoughts beyond the reaches of our souls?
Say, why is this? Wherefore? What should we do?
(*Hamlet*, Act I, Scene 4)

By now it should be clear that the argument of this book is an extended ghost story. If I have been successful, I will have succeeded in conveying the uncanny effects of post-closure cold fusion as I experienced them in my fieldwork. My goal has been not simply to trace the ways that cold fusion research manages to survive; I have also tried to show that cold fusion's existence is contingent on the particular relations between cold fusion researchers and mainstream scientific culture. To understand the situation, consider this counterfactual hypothesis: if cold fusion had not "died," none of the things I have described would be features of the current research world. This point can be grasped only in the importantly asymmetrical glance, because if we cannot allow cold fusion to be dead, we will not be able to see the myriad ways in which it continues to survive, always in the shadow of the life it desires but does not have. The predicament of the ghost is to be caught positively (as a stable entity) between life and death, but always in relation to life (legitimacy) on one side and death (illegitimacy) on the other.

I leave open the task of further exploring this relation in both directions. Through the continued analysis of undead science, we can learn more about living sciences on the one hand and dead sciences on the other. Indeed, just as cold fusion's survival is a condition of the configuration of mainstream science, so the configuration of mainstream science is conditioned through its specific relations with undead sciences like cold fusion research. This will become obvious if cold fusion researchers actually manage to reopen the case.[13] But even if they do not, it would be unwise to ignore the concrete effects of cold fusion research on mainstream scientific practice, just as it would be unwise to ignore the relation between Newton's interest in alchemy and his natural-philosophical work (Westfall 1980).[14]

Finally, I want to suggest that there are even stronger epistemological consequences at stake once we start taking account of the practices of the undead. Mainstream scientific culture produces truth not only because some knowledge claims are made into facts through the actions and associations of certain people and things, but also because in the same motion some other knowledge claims are made into artifacts or fictions.[15] These artifacts do not just disappear, because (and here I am being symmetrical) wherever they go, some other people and things must go also, and people and things don't usually just disappear (although it is important to point out that under certain political conditions they can literally disappear). In the end, the actions and associations of these "artifactualized" people and things cannot be either asymmetrically discounted or symmetrically relativized, because in point of fact, their existence circumscribes and is circumscribed by what gets to count as truth and who gets to count as a truth-teller in our culture. Thus, to ignore the operations of the undead is simply to neglect the fact that they will be back to haunt us.

# Notes

## *Chapter 1* **Science Studies and Second Sight**

1. From 1993 to 1996, I worked off and on as a volunteer research assistant for RF, a materials engineer doing cold fusion experiments. I have elected to make use of false initials for the names of the principal actors involved in order to preserve anonymity.
2. In electrolysis experiments like ours where gases are allowed to escape into the air rather than putting the apparatus in a fume hood, using lithium can be dangerous.
3. It has been difficult to gauge exactly how many scientists became involved in cold fusion research during the summer of 1989. Anecdotal evidence suggests that some kind of investigation took place at almost every major university and research laboratory around the world. The Institute for Scientific Information (ISI) ranked cold fusion the "hottest" field of research in 1989 based on the number of papers published around a single core subject. Lower-ranking fields included high-temperature superconductivity, HIV transmission, identification of the cystic fibrosis gene, and the atmospheric chemistry of Antarctica. See ISI (1990a).
4. E-mail interview with a chemist, April 12, 1994.
5. University of Illinois materials scientist Howard Birnbaum, quoted in Bishop (1996).
6. The strongest formulation of this perspective can be found in actor-network approaches to the study of science, especially Latour (1993). I will discuss the conceptualization of knowledge as social-material convergence in chapter 7.
7. On alternative metaphysics for science studies and the notion of variable ontologies, see, for example, Latour (1996) and Cussins (1996).
8. I want to thank Jill Didur for the idea of using the concept of hauntology here. The term is borrowed, with some sympathetic alteration, from Derrida (1994).
9. In this sense the argument of this book is meant to share an affinity with the critical science studies of Haraway (1992, 1994).
10. Although they disagree on many other points: see Bloor (1999) and Latour (1999).
11. See Wallis and Morley (1976), Wallis (1979), Barnes and Shapin (1979), Nowotny and Rose (1979), and Collins and Pinch (1982).

12. I leave out consideration of important work in the study of technical controversies (Nelkin 1979; Markle and Peterson 1981) or science-based controversies (Brante, Fuller, and Lynch 1993).
13. See especially Barnes, Bloor, and Henry (1996) and Collins and Pinch (1993).
14. Popular books include Peat (1989), Close (1991a), Mallove (1991), Huizenga (1992), and Taubes (1993); books for a more specialized audience include Hoffman (1995) and Beaudette (2000). More sociological and historical articles include Lewenstein (1992a), Gieryn (1992), McAllister (1992), Sullivan (1994), Pinch (1994), Gross (1995), Taylor (1996), and Toumey (1996).
15. The Cornell Cold Fusion Archive (CCFA) was developed by Bruce Lewenstein and Tom Gieryn. The archive is maintained in the Rare and Manuscript Collections, Carl A. Kroch Library, Cornell University, collection no. 4451.

## *Chapter 2* **The Birth of Cold Fusion**

1. For one such vision see Stirling (1991).
2. It is not clear whether the notion of the core set is as useful in the case of science-based controversies (see Brante 1993), where participants may include members of government, business, social movements, and so on, since such cases are sociologically speaking much more complex. In general, the idea of the core set becomes useless if it is considered to be extendable to all actors in a controversy.
3. An interesting exception relevant to this discussion is Michael and Birke (1994).
4. Transcript from a videotape of the press conference, Cornell Cold Fusion Archive (CCFA #4451).
5. Transcript, which is the source of subsequent quotations from the press conference.
6. Amongst other things, as explained in chapter 1, electrolysis is the most common procedure for separating water into its molecular constituents, hydrogen and oxygen. It is a basic chemical technique taught in high school chemistry classes.
7. See Gieryn (1992) for a more detailed discussion and analysis of the narrative dimension of the public representations of cold fusion. For a more general discussion of narrative in science journalism see Curtis (1994).
8. From "Polywater, Fusion and the Popular Press," unpublished manuscript. This manuscript was submitted for publication to a Canadian news magazine on March 28, 1989. It was rejected for publication by the editors of the magazine.
9. It is important to note that Fleischmann and Pons had in fact submitted a paper to the *Journal of Electroanalytical Chemistry* prior to the press conference, which was published as a non-peer-reviewed preliminary note on April 10, but by then attempts at experimental replication were already well under way.
10. For a discussion of the development of patenting culture in science, see Packer and Webster (1996).
11. The best account to date of what transpired can be found in Close (1991a), and a similar version in Taubes (1993). Also see Lewenstein (1992b). There is still much disagreement about the interpretation of the events leading up to March 23, 1989, but in retrospect most of the actors involved agree that holding the press conference was a bad idea.
12. Gieryn borrows the term from Secord's (1989) work on a nineteenth-century spontaneous-generation controversy, which also received a great deal of popular attention. For another comparison using the same case, see Klotz and Katz (1991).
13. Interview with an electrochemist, July 1, 1994.
14. Interview with Charles Barnes conducted by Douglas Smith, May 15, 1989, CCFA, box 9.
15. Quoted in Smith (1989, 6).

16. In a series of taped interviews housed at the CCFA, scientists were asked how they learned about cold fusion and began their experiments. The apparent contradiction between negative perceptions of media reports and positive use in the design of early experiments is nearly uniform amongst those interviewed.
17. From "Polywater, Fusion and the Popular Press"; see note 8.
18. For a discussion of this, see also Mulkay and Gilbert (1986).
19. Interview with a French engineer, May 31, 1995.
20. See Hilgartner (1990) for a discussion of this. The communication pathway does vary somewhat, depending on the degree to which particular newspaper science writers are integrated into the scientific community. They may, on some occasions, receive information directly from the original scientists, but even then usually only after a journal article has been published.
21. See Collins (1988). Media representations of cold fusion cannot be understood as public displays of virtuosity in the sense that Collins suggests.
22. The term "science worlds" is borrowed from Howard Becker (1982) and related literature using the "social worlds" framework (e.g., Star and Griesemer 1989, Clarke 1990). I find the concept of science worlds useful for talking about the myriad of actors other than professional scientists whose participation is intrinsic to the production of natural-scientific knowledge (see chapter 6).
23. For a copy of the press release, see Huizenga (1992, 237–239). For a detailed account of the press conference, see Gieryn (1999, 187–203).
24. ABC news correspondent Jim Slade on March 23, 1989.
25. Charles Martin, quoted in Taubes (1993, 110).
26. Other common terms include "cold nuclear fusion" and "test-tube fusion."
27. While the majority of fusion processes involve atomic reactions occurring at very high temperatures and pressures, muon-catalyzed fusion occurs under less extreme conditions when a muon replaces an electron in a closer orbit around a hydrogen nucleus (which consists of one proton). This process produces a short-lived hybrid called muonic hydrogen with an overall neutral charge, making the possibility of fusion more likely since there is less repulsion between the otherwise positively charged nuclei. Due to the instability of muonic hydrogen, however, the rate of muon-catalyzed fusion is so low as to be unusable in commercial energy production.
28. See Jones et al. (1989). The BYU experiments were originally designed to explore the possibility of naturally occurring fusion in the high-pressure regions in the mantle of the earth, a process that worked differently from muon-catalyzed fusion and that Jones and a coauthor decided to call piezo-nuclear fusion.
29. Interview with a nuclear physicist, March 24, 1996.
30. Huizenga, for instance, attempts to make this point in the introduction to his book. There is no doubt that the results in the two sets of experiments were seen differently by Jones on one side and Fleischmann and Pons on the other, but this does not account for the ways in which the results were represented and interpreted by others.
31. See Gieryn (1999, chapter 4) for an interpretation along these lines.
32. This was despite an apparent agreement between BYU and the University of Utah noted above that papers would be submitted simultaneously to *Nature*.
33. Private e-mail from an electrochemist, March 20, 1996.
34. Interview with a physicist, April 24, 1996.
35. "Static" calorimeters measure temperature difference in situ, while "flow" calorimeters measure the temperature of the electrolyte before it enters the cell and after it leaves. It is generally thought that flow calorimetry is more precise since it eliminates possible artifacts.

36. "Break-even" is a term borrowed from conventional fusion research. In this case, the paper is referring to "scientific break-even" the point at which the total power input (usually measured in watts) equals the total power output. A break-even value over 100%, or $Q > 1$, indicates the production of excess power. A similar relation is determined for the production of heat (which is measured in joules) in the cell, and it is important to note that these values are not equivalent.
37. Possible chemical reactions include the recombination of oxygen and deuterium in the electrolyte, a redox reaction between oxygen and palladium, and reactions of oxygen or deuterium with impurities in the electrolyte.
38. Only nuclei with an atomic mass less than 60 will produce energy when fused; heavier nuclei tend to absorb energy instead.
39. This is to say that branching ratios are not simply a theoretical construct, but are instrumental, for instance, in determining the thickness and composition of fusion reactor walls.
40. For a criticism of the Walling-Simons theory, see Huizenga (1992, 37–38).
41. See Peat (1989, chapter 3) for an explanation of this process.
42. This process is known as "quantum tunneling."
43. See, for example, Koonin and Nauenberg (1989).
44. See Huizenga (1992), 153.
45. Interview with an electrochemist, July 1, 1994.
46. The idea here is a simplified version of the notion of interpellation used routinely in structuralist semiotics.
47. See Sismondo (1996) and Hacking (1999) for an account of the many forms this kind of construction can take.

## *Chapter 3* **The Cold Fusion Controversy**

1. The prevailing explanation for superconductivity was once known as the BCS theory. This theory, first published in 1957 by John Beardon, Leon Cooper, and Robert Schrieffer (BCS), set a cap of 23 degrees Kelvin on the critical temperature for superconductivity, and this seemed to predict the experience with experiments very well. Müller and Bednorz's innovation, after a long period of trial-and-error experimentation, was to use ceramic materials rather than metals. With some difficultly they produced a lanthium–barium–copper-oxide compound that was superconducting at 30 K.
2. The most important innovation was made by Paul Chu's group at the University of Houston in 1987. Chu's group used yttrium instead of lanthium and produced a material that was superconducting at 90 K (as opposed to Müller and Bednorz's 30 K). Chu's innovation made $HT_c$ research important commercially, and as a consequence the field mushroomed. For an excellent account of Chu's work written by a participant, see Hazen (1988).
3. I had wondered about this. Did Fleischmann expect cold fusion to be replicated as simply as HTc was? In a conversation with me, Fleischmann indicated that he had not expected replication to be easy, but he had also not expected that the criticism of his work would be so fierce either.
4. These criteria are discussed in Hazen (1988, 23). The Meissner effect provides the clearest demonstration of superconductivity; when cooled to its "critical temperature," a permanent magnet can be made to "float" above a piece of superconducting material.
5. The situation is in fact a little more complex than I have described it here. The Meissner effect, for instance, is exhibited slightly differently in pure metal (type 1) superconductors than in metallic alloys (type 2). In most instances, however,

the recognition of superconductivity in materials, even in new compounds, is a routine and unproblematic procedure.

6. This statement reflects the general definition of closure for analysts employing EPOR or actor-network theory in the study of controversies. In applying this definition to both approaches, I have intentionally blurred a crucial distinction between closure as agreement in belief and closure as agreement in practice. The reason for this is that in terms of the argument presented here, agreement in belief can be subsumed under agreement in practice. For a broader range of arguments relevant to the discussion of closure see Englehardt and Caplan (1987), Galison (1987), Baigrie and Hattiangadi (1992), Hard (1994), and Kim (1996).
7. Nathan Lewis in Smith (1989, 10).
8. Interview with an electrochemist, July 1, 1996.
9. J. Appleby, Y. Kim, C. Martin, O. Murphy, and S. Srinivasan, "Anomalous Calorimetric Results during Long-Term Evolution of Deuterium on Palladium from Lithium Deuteroxide Electrolyte," unpublished manuscript, Cornell Cold Fusion Archive, box 7.
10. While Martin's group was still left with some 10 to 20% excess, it was unable to reproduce the effect after forty more experiments and discontinued its CF research by the end of 1989.
11. Furth was not the first to propose this; the issue had been raised at the talk Fleischmann gave at the Harwell Nuclear Research Laboratory in England on March 28. A number of critics had expressed concern that Fleischmann and Pons did not seem to have considered this important control, or if they did, they did not report it.
12. Statement of Harold P. Furth before House Committee on Science, Space, and Technology, April 26, 1989; CCFA, box 8.
13. Quoted in Pool (1989).
14. Fleischmann has remained silently skeptical of the claims of fusion in light-water experiments, and the issue remains a source of some dispute amongst cold fusion researchers today (see chapter 5).
15. Statement of Robert A. Huggins before House Committee on Science, Space, and Technology, April 26, 1989; CCFA, box 8.
16. In terms of understanding closure, interviews and private conversations become less illuminating. Actors can always find ways of disagreeing or agreeing with a claim, as Collins (1975, 1981c) aptly shows, but analysts must balance what actors could do with what they actually end up doing. The methodological problem here concerns identifying which statements actually get put into play, a problem sometimes complicated by interviews.
17. This interest was most noticeable in Utah, where "fusion fever" took the form of daily news reports on television and radio, songs and T-shirts, and mugs and baseball caps (a variety of such paraphernalia is housed in the Cornell Cold Fusion Archive).
18. See Smith (1989) for an account of the Caltech effort. Smith's article is a valuable resource, as participants are quoted at length from interviews performed in May 1989.
19. Quoted in Taubes (1993, 123).
20. Quoted in Taubes (1993, 189).
21. Quoted in *Chemistry and Engineering News*, May 8, 1989, 4.
22. Members of Lewis's group interviewed by Smith (1989) suggested that his change of heart was prompted by Fleischmann and Pons's appearance in front of the House Committee on Space, Science, and Technology on April 26. His feeling was that the good will of other scientists had been abused by this act.

23. Interview with an electrochemist, July 1, 1995.
24. This interpretation is supported by Pinch's (1996) interesting analysis of the rhetorical use of humor in some of the presentations at the APS meeting.
25. E-mail from DOE scientist, April 16, 1989.
26. E-mail from IBM scientist, April 19, 1989.
27. Quoted in the *Salt Lake City Tribune*, May 10, 1989.
28. Quoted in the *Salt Lake City Tribune*, May 10, 1989.
29. Caltech team member Michael Sailor, quoted in Smith (1989, 8).
30. E-mail message to an electronic discussion list on anomalous energy phenomena, August 17, 1996.
31. Nathan Lewis on the PBS radio program *Science Journal*, May 4, 1989.
32. Quoted in Taubes (1993, 298).
33. Peter Bond, quoted in Stipp (1989).
34. See Close (1991a, 279–288, 1991b).
35. A copy of the letter from Triggs to Salamon was posted with the permission of Salamon on the Usenet cold fusion discussion group sci.physics.fusion, June 6, 1990.
36. Salamon did not retract his results, and Pons never followed through on his threat to sue.
37. Interview with an electrochemist, July 1, 1995.
38. Excerpt from a personal letter from a physicist to the director of the Institute for Theoretical Physics at UCLA, May 5, 1989.
39. Dieter Britz, "Current CNF Status," message posted to sci.physics.fusion, October 9, 1991.
40. Quoted in Huizenga (1992, 89).
41. Quoted in Huizenga (1992, 90).
42. Interview with an electrochemist, July 1, 1995.
43. Interview with a materials engineer, November 11, 1996.
44. Interview with a physicist, March 4, 1996.
45. Interview with an engineer, October 6, 1995.
46. Interview with a nuclear engineer at a private company, October 6, 1995.
47. Interview with a nuclear engineer at a private company, October 6, 1995.
48. Interview with a chemist, July 7, 1996.
49. Reviewer's comments on a paper submitted to a chemistry journal by a cold fusion researcher in 1994.

## *Chapter 4* **The Pallor of Death**

1. It is worth noting that Eugene Mallove, the author of one of the first popular books on the cold fusion controversy (Mallove 1991) and the founder-editor of *Infinite Energy: The Magazine of New Energy Technology*, served as a technical consultant for the film.
2. The solution draws on the traditional plot device of a "secret formula," which is stored in Emma's bra and is later stolen by Simon Templar, who is working for the Russian crime lord. (He later undergoes a moral conversion after falling in love with Emma.) The formula has no historical correlate to the actual controversy, but the plot device is interesting vis-à-vis public perceptions of science as an essentially algorithmic practice (Collins 1985, 20–23).
3. This is not an insignificant point. *The Saint* suits preferences of students in engineering and the sciences for action films, and given the degree of marketing the film received, it was unlikely to go by unnoticed (by younger male scientists especially).
4. This is in spite of the fact that one of the most enthusiastic audiences for the film

was the cold fusion researchers themselves. The reason why this is so will become evident in chapter 5.

5. See also Mulkay (1976) and Prelli (1989).
6. This issue is also discussed in the context of the public policing of science in Shapin (1993).
7. From an e-mail interview with a chemist, November 20, 1994.
8. A written version of Langmuir's lecture was transcribed in 1968 from what was apparently a bad microgroove disk recording (report no. 68-C–035, edited by R. N. Hall, General Electric Research and Development Center, Schenectady N.Y., April 1968). The lecture started to circulate widely only after 1985, when it was published in *Speculations in Science and Technology*, and 1989, when it was published in *Physics Today*, although it was part of the oral culture of American scientists before then.
9. For further discussion of these episodes, see Nye (1980) and Ashmore (1993) for the case of N-rays, and Picart (1994), Schiff (1994), and Fadlon and Lewin-Epstein (1997) for the case of water memory.
10. See Taubes (1993, 262–264)
11. To put it another way, I would suggest that the easy circulation of pathology talk both indicates and constitutes the process of closure in the cold fusion case.
12. Quoted in Taubes (1993, 265).
13. See Beaudette (2000, 83–86) for a critical review of this article.
14. Morrison was a participant in the informal network of fusion researchers. See chapter 3.
15. Morrison was perhaps the only skeptic left from the original core set of 1989 who regularly attended conferences on cold fusion, up to his death in 2001. While there may be some evidence that his position had softened over the years, he maintained that his primary interest in the case was to develop a better understanding of pathological science.
16. This suggestion was disavowed by members of the EPRI, who explained that during the visits research funding was not in question.
17. E. Fry, J. B. Natowitz, and J. Poston, "Report of the Cold Fusion Review Panel at Texas A&M University," October 15, 1990, CFFA, box 9.
18. Pace's evidence for fraud is unclear. He may have been referring to the implicit accusation of deception made by Richard Petrasso et al. in their *Nature* article on the gamma spectrum of Fleischmann and Pons, which was also discussed by Close (1991a) in his book published just before Pace's article.
19. Quoted in a translation of an Italian court transcript by Luigi Bianchi, York University, November 10, 1995.
20. Italian court transcript, translated by Bianchi.
21. "Report of the Cold Fusion Review Panel at Texas A&M University," 16.
22. Quoted in Taubes (1993, 403).
23. Mallove (2000) and letters in CCFA, box 9, folder 31. Similar difficulties have been experienced by other students working on cold fusion, most recently in Japan and Italy.
24. Interview with a physicist, March 4 1996.
25. I am paraphrasing a comment that has been made to me on several occasions by scientists (mostly physicists), some of whom are knowledgeable about science studies, some not.
26. Private e-mail from a sociologist of science, January 31, 1994.
27. See, for instance, Thomsen (1999).
28. See "Macromedia ColdFusion 5" (http://www.macromedia.com/software/coldfusion/) and "Cold Fusion Foods" (http://www.coldfusionfoods.com/).

29. Or perhaps not; Pepsi-Max was not particularly successful, and given its absence from store shelves today, it may actually parallel the story of cold fusion in more ways than I have suggested here.
30. Other cases considered in the book include "natural" and new age medicine, the supposed carcinogenic effects of power lines, and UFOs.
31. Frank Close, public lecture at the University of Toronto, February 20, 1994.

## *Chapter 5* **The Afterlife of Cold Fusion**

1. See Fleischmann and Pons (1994b).
2. See Amato (1993) and a reply from CF researcher Richard A. Oriani (1993). See also Brown (1993). A more positive response can be found in White (1993).
3. Interview with a member of the editorial board of *Physics Letters* A, January 16, 1994.
4. According to some skeptics, there are still unaccounted-for artifactual sources of heat in these experiments. One possibility is that the input energy is stored in the lattice over a period of time and then released later. Fleischmann and Pons argued against this, saying that their calculations take this into account. The problem is that these energy balance calculations do not have the same rhetorical poignancy as the heat-after-death demonstration.
5. I attended the conference where Fleischmann and Pons described their heat-after-death results. Two researchers working in an industrial R&D laboratory told me over lunch that they had been unable to see any CF effects in their experiments and that the conference was their last-ditch effort to learn something that would make a difference before they gave up.
6. Message posted to sci.physics.fusion newsgroup, May 20, 1993.
7. The kind of analysis being attempted both by me and by Harry Collins, and any disagreement between us, should be viewed as being different from the conflicts that scientists themselves would engage in on the same topic.
8. These meetings are supplemented by a series of smaller national and international workshops. Regular gatherings of CF researchers occur in Japan (under the auspices of the Japanese Cold Fusion Society), Italy, and Russia. In the United States, researchers have met in a variety of forums, including conferences on new energy technologies and panels at meetings of the American Physical Society and the Electrochemistry Society.
9. For an overview of some of this work, see Sproull and Kiesler (1991), Calhoun (1992), and Jones (1995).
10. It is important to understand these two features as being related. Even if a scientist does not answer an interlocutor's questions well, the failing may be excused if the scientist is deemed to be a good or honest person ("He is under a lot of pressure lately," or "He couldn't tell me anything because of patent restrictions").
11. Robert Park referred on a television news broadcast to the Salt Lake City meeting as a "séance of a hardy band of remaining true believers" (see Bapis 1990).
12. Quoted in Mallove (1991, 215).
13. Quoted in Bapis (1990, 6).
14. For a related discussion of resistance identity in social movements, see Gamson (1997), and in the context of a scientific research movement, see Epstein (1993) and Indyk and Rier (1993).
15. Interview with an electrochemist, July 1, 1995.
16. Interview with a chemical engineer, February 21, 1994.
17. My experience as a participant observer was very important in drawing out respondents during interviews. Respondents who were professional scientists were often

guarded about using the term cold fusion until I talked at length about my own experiences in conducting experiments, attending conferences, and talking to other researchers. Some respondents also needed reassurance that I was not a reporter, or that I would preserve their anonymity.

18. Erving Goffman's (1969) work on the performance of identity is also relevant here. There is a sense that the researchers quoted here are drawing a distinction between an "onstage" and a "backstage" identity. There is the label they use in public, on paper, and for their employers (i.e., the label they are publicly accountable for) and the label they use in private, for themselves or with fellow CF researchers. Conceived of in these terms, the relatively public event of a "cold fusion" conference is not only a vocal expression of resistance identity (and a kind of identity politics) in Castells's terms, but also a constitutive backstage where CF researchers can "relax" their performance in a context less hostile to their work (Mukerji and Simon 1998).
19. E-mail interview with a CF researcher, October 14, 1993.
20. Interview with a theoretical physicist, March 4, 1996.
21. See, for example, William Beatty, "Symptoms of Pathological Skepticism," website http://www.eskimo.com/~billb/pathsk2.txt (1996).
22. From a letter written by a chemist at a large research laboratory in India, April 8, 1991.
23. From the same letter. In a conversation in December 1994, the author of the letter told me that the director of his laboratory withdrew funding for cold fusion research in 1990 and that he has been unsuccessful in convincing senior administrators to review either his own or others' experimental evidence in the hope of reviving his research project.
24. Technova, Inc. was established by Minuro Toyoda as a consortium for funding a variety of speculative research projects. Together with IMRA, Technova was part of Toyoda's plan to establish international research and development links. After meeting with Toyoda, Fleischmann and Pons applied to Technova for funding in 1990.
25. The device was called "ICARUS," which stands for isoperibolic calorimetry research and utility system. ICARUS 1 was designed primarily as a data acquisition system that could demonstrate excess-heat effects in working CF cells. Later designs were supposed to integrate working cells with data acquisition in an effort to devise a "black box" demonstration device that could be transported and studied by different groups. At the present time no such device has been released.
26. As I work on the revisions for this chapter on August 31, 2001, I am listening to the National Public Radio program *Science Friday*, which features a report on the production of anomalous tritium in a cold fusion experiment at the Stanford Research Institute.
27. McKubre was able to produce excess-heat results in 1989, and although his work was not featured in any media reporting at the time, he was one of the scientists consulted by the DOE ERAB panel during the summer of 1989.
28. Quoted in Beaudette (2000, 189).
29. Interestingly, Beaudette (2000, 284) reports that McKubre was visited in his laboratory by a number of scientists, amongst them Nathan Lewis (the electrochemist responsible for the damaging critique of Fleischmann and Pons in May 1989), who did not point out anything particularly wrong with McKubre's measurements.
30. In 1992, Akito Takahashi at Osaka National University reported excess heat and low-level neutron emissions at the same time. As with Fleischmann and Pons's 1989 results, the neutrons were too few to account for the excess heat by convention fusion, but this research has not been followed up (see Beaudette 2000, 220).

31. Other important results come from Edmund and Carol Talcott Storms, also of Los Alamos.
32. In one talk I attended at the Engineering School at the University of California, San Diego, Miles spent a great deal of time discussing his procedures for eliminating sources of contamination. Members of the audience who were not CF researchers seemed to think he had done a good job, but others like Steven Jones have been more critical.
33. Gozzi has also produced important results by measuring with an "in-line mass spectrometer" the production of He–4 in real time in a cell producing excess energy.
34. Many of these experiments were reported at ICCF–3 in Nagoya and ICCF–4 in Maui. The first publication was Mills and Kneizys (1991). Mills has since created a company called Blacklight Power Inc. (see http://www.blacklightpower.com) and is trying to develop a heating device.
35. Another change was the use of potassium carbonate instead of lithium as a catalyst in the electrolyte.
36. For instance, Douglas Morrison, "Cold Fusion Update No. 7," e-mail message posted to sci.physics.fusion, November 27, 1993.
37. Interview with an electrochemist, July 1, 1995.
38. See Miley and Patterson (1996).
39. Petition signed by twenty-three faculty and sent to the provost of Texas A&M University, quoted in Mallove (1994).
40. Interview with a CF researcher, March 2, 1996.
41. Important ideas for the argument of this chapter come from Gooding, Pinch, and Schaffer (1989), Pickering (1992), and Buchwald (1995).
42. I want to distinguish the analysis here from work on cognitive dissonance such as that by Festinger, Riecken, and Schachter (1956), but the beliefs of the members of doomsday cults might be enabled and constrained by material practices in ways that are not much different from those of cold fusion researchers or any other kinds of scientists.

## *Chapter 6* **Tales from the Crypt**

1. Park (2000) identifies the ABC news report of Patterson's claims as exemplary of bad science (and bad science reporting).
2. Only a few of these were ever sold, as CETI decided it did not want to expend the resources to support the users of the RIFEX system. Of the few users who obtained a device, none reported measuring excess heat with it.
3. See, for instance, Park (2000).
4. Examples of such actors include George Miley, who while also doing CF experiments was until recently the editor of the *Journal of Fusion Technology*. Miley was responsible for devoting a small section of the journal to CF research. Other actors include Johnson-Matthey, Inc., which has supplied the greatest number of active palladium samples in successful CF experiments (but no longer does so), and Eugene Mallove, who publishes the magazine *Infinite Energy* and supports a small laboratory for testing anomalous-energy devices.
5. Interview with a physicist, March 4, 1996.
6. See chapter 1, note 1. In this section I present only a small portion of my fieldwork data as they pertain to the issue of the ghostly qualities of core-group work on CF. This necessitates a slight shift in narrative style.
7. Interview with an engineer, August 6, 1995.
8. I do not mean to inflate the value of my participation. The experiments certainly could have been performed without me, but my experience testifies to the impor-

tance of general busywork associated with most experimentation. Since my departure from the project, RF retains an interest in cold fusion, but few new experiments have been performed.

9. Interview with RF, November 9, 1996.
10. The uniqueness of our data attracted the interest of a retired scientist and CF researcher working at a nearby institution. We eventually arranged to do IR scans using his CF cells, and he incorporated some of our work in a publication.
11. E-mail comment reproduced in *Cold Fusion Times*, 1, 4 (winter 1994): 3.
12. Interview with a retired systems engineer, September 16, 1996.
13. Some critics of cold fusion argue that Miley's editorial policies also made it impossible for any papers critical of cold fusion to be published. At least one scientist complained to me that his critical paper had received such unfair treatment by reviewers that he had to send it to a different journal, where it would be less likely to be read by CF researchers.
14. The sample issue is vol. 3, issue 18 (March 1998).
15. Interview with an engineer, August 6, 1995.
16. Interview with an electrochemist, July 1, 1995.
17. NEDO is a semiprivate corporation that was established by MITI to nationalize Japan's coal-mining industry and carry out R&D on new forms of energy. Cold fusion research amounts to a very small portion of NEDO's budget.
18. "NEDO Bites the Dust," e-mail message to cold fusion discussion group, March 18, 1997.
19. This is a basic strategy for CF researchers seeking grants and patents for their work. In the United States, the strategy seems to have met with little success. In the case of patenting, the very claims that make the research patentable in the first place are the ones that identify it as being related to cold fusion. For legal reasons, most of these patents must mention Fleischmann and Pons's 1989 paper, and this serves to tip off the reviewing patent officer.

### *Chapter 7* **A Hauntology for the Technoscientific Afterlife**

1. A different view is taken by Scott, Richards, and Martin (1990), amongst others. See Collins (1991) for a reply.
2. By way of comparison, consider Gerold Holton's account of the Millikan-Erenhaft controversy as discussed by Barnes, Bloor, and Henry (1996, 35–40).
3. For a perspective along these lines, see Collins (1999). Papers of rejected scientists may certainly make it into mainstream journals, but the absence of any response from other scientists results in what Collins calls the "interpretive death" of the paper.
4. Early controversy studies can be read as staking a claim for the possibility of a sociology of scientific knowledge against skeptical criticism from rationalist philosophers, historians, and sociologists of science. And of course establishing stable sociological knowledge about science can be seen reflexively as a social achievement (or failure) in defining an SSK community.
5. Note that this does not necessarily imply that the "data" have no role in their interpretation, only that the qualities of the data alone cannot determine the correct interpretation.
6. At one time the philosopher Larry Laudan (1977) argued that this was all that the sociology of scientific knowledge would be good for. Here the debate is between Laudan and Bloor (1991).
7. See Pinch (1986b) for a critical review of Rudwick (1985) along these lines.
8. See also Latour (1988b, 33–35), and for a critique, see Carlson and Gorman (1992).

9. The agonistic metaphors of actor-network theory are most prominent in Latour (1988a), in which Louis Pasteur's "battles" with his opponents take on Napoleonic proportions. For an insightful critique of actor-network theory, see Brown and Lee (1994).
10. This is the concluding comment from John Maddox in the TV documentary "Confusion in a Jar" which first aired on the BBC *Horizon* series on March 26, 1990.
11. Interview with a chemist, May 29, 1996.
12. Note that it is not clear whether the Azande or members of cults can be so easily made to seem different either.
13. In this event, accounting for the time cold fusion spent as "dead" will be crucial in explaining how it managed to be resurrected, especially since from an asymmetrical historical perspective it will appear as if cold fusion had always been alive.
14. Since most cold fusion researchers are also professional scientists (though they may be "marked"), it stands to reason that a definite relation will obtain between any given scientist's cold fusion practices on the one hand and that scientist's "normal" scientific practice on the other. At least one research group working on superconductivity, for instance, has linked its work to cold fusion. See Johnson and Clougherty (1989).
15. Which is to say that truth and not-truth (falsity) are the twin productions of the same social and material processes.

# *Works Cited*


Allison, P. D. 1979. "Experimental Parapsychology as a Rejected Science." In *On the Margins of Science: The Social Construction of Rejected Knowledge*, edited by R. Wallis, pp. 271–292. Keele, U.K.: University of Keele Press.

Amato, I. 1993. "Pons and Fleischmann Redux?" *Science* 260: 895.

Ashmore, M. 1993. "The Theatre of the Blind: Starring a Promethean Prankster, a Phony Phenomenon, a Prism, a Pocket, and a Piece of Wood." *Social Studies of Science* 23:67–106.

Baigrie, B., and Hattiangadi, J. N. 1992. "On Consensus and Stability in Science." *British Journal for the Philosophy of Science* 43:435–458.

Bapis, J. 1990. "Pons, Fleischmann Hope Fusion Furor Will Give Way to Solid New Research." *University of Utah Review* 23 (April/May): 5–7.

Barnes, B. 1982. *T. S. Kuhn and Social Science*. London: Macmillan.

Barnes, B., D. Bloor, and J. Henry. 1996. *Scientific Knowledge: A Sociological Analysis*. Chicago: University of Chicago Press.

Barnes, B. and Shapin, S. 1979. *Natural Order: Historical Studies of Scientific Culture*. London: Sage.

Beaudette, C. 2000. *Excess Heat: Why Cold Fusion Research Prevailed*. South Bristol, Maine: Oak Grove Press.

Becker, H. S. 1982. *Art Worlds*. Berkeley: University of California Press.

Bishop, J. 1989a. "Research in Utah to Announce a Development in Fusion Energy." *Wall Street Journal*, March 23, B1.

Bishop, J. 1989b. "Scientist Sticks to Claimed Test-Tube Fusion Advance." *Wall Street Journal*, March 27, B3.

Bishop, J. 1996. "A Bottle Rekindles Scientific Debate about the Possibility of Cold Fusion." *Wall Street Journal*, January 29, A9.

Bishop, J., and K. Wells. 1989. "Taming H-Bombs? Utah Scientists Claim Breakthrough in Quest for Fusion Energy." *Wall Street Journal*, March 24, A1.

Bloor, D. 1991. *Knowledge and Social Imagery*. 2nd edition. Chicago: University of Chicago Press.

Bloor, D. 1999. "Anti-Latour." *Studies in the History and Philosophy of Science* 31:81–112.

Bockris, J.O.M. 2000. "Accountability and Academic Freedom: The Battle Concerning

Research on Cold Fusion at Texas A&M University." *Accountability in Research* 9:103–119.

Botting, F. 1996. *Gothic*. New York: Routledge.

Brante, T. 1993. "Reasons for Studying Scientific and Science-Based Controversies." In *Controversial Science: From Content to Contention*, edited by T. Brante, S. Fuller, and W. Lynch, pp. 177–192. Albany: State University of New York Press.

Brante, T., S. Fuller, and W. Lynch, eds. 1993. *Controversial Science: From Content to Contention*. Albany: State University of New York Press.

Britz, D. N.d. "Cold Fusion Bibliography." Website at http://www.kemi.aau.dk/~britz/fusion/fusion.html.

Brown, N., and S. Lee. 1994. "Otherness and the Actor Network: The Undiscovered Continent." *American Behavioral Scientist* 37: 772–790.

Brown, W. 1993. "Frosty Reception Greets Cold Fusion Figures." *New Scientist* 1871:6.

Buchwald, J. 1995. *Scientific Practice: Theories and Stories of Doing Physics*. Chicago: University of Chicago Press.

Calhoun, C. 1992. "The Infrastructure of Modernity: Indirect Social Relationships, Information Technology, and Social Integration." In *Social Change and Modernity*, edited by H. Haferkamp and N. Smelser, pp. 205–236. Berkeley: University of California Press.

Calhoun, C. 1994. *Social Theory and the Politics of Identity*. Oxford: Blackwell.

Callon, M. 1980. "Struggles and Negotiations to Define What Is Problematic and What Is Not: The Socio-Logic of Translation." In *The Social Process of Scientific Investigation*, edited by K. Knorr, R. Krohn, and R. Whitley, pp. 197–219. Dordrecht: D. Riedel.

Callon, M. 1986. "Some Elements of a Sociology of Translation: Domestication of the Scallops and the Fishermen of St. Brieuc Bay." In *Power, Action and Belief*, edited by J. Law, pp. 196–229. London: Routledge and Kegan Paul.

Callon, M., and B. Latour. 1992. "Don't Throw the Baby Out with the Bath School! A Reply to Collins and Yearly." In *Science as Practice and Culture*, edited by A. Pickering, pp. 343–368. Chicago: University of Chicago Press.

Carlson, W. B., and M. E. Gorman. 1992. "Socio-Technical Graphs and Cognitive Maps: A Response." *Social Studies of Science* 22:81–91.

Castells, M. 1996. *The Rise of the Network Society*. Oxford: Blackwell.

Clarke, A. 1990. "A Social Worlds Research Adventure: The Case of Reproductive Science." In *Theories of Science in Society*, edited by S. Cozzens and T. Gieryn, pp. 23–50. Bloomington: Indiana University Press.

Clarke, A. 1991. "Social Worlds/Arenas Theory as Organizational Theory." In *Social Organization and Social Process: Essays in Honor of Anselm L. Strauss*, edited by D. Maines, pp. 119–158. Hawthorn, N.Y.: Aldine de Gruyter.

Close, F. 1991a. *Too Hot to Handle: The Race for Cold Fusion*. Princeton: Princeton University Press.

Close, F. 1991b. "Cold Fusion I: The Discovery That Never Was." *New Scientist* 129:46–48.

Collins, H. M. 1975. "The Seven Sexes: A Study in the Sociology of a Phenomenon, or the Replication of an Experiment in Physics." *Sociology* 9:205–224.

Collins, H. M., and R. G. Harrison. 1975. "Building a TEA Laser: The Caprices of Communication." Social Studies of Science 5:441–450.

Collins, H. M. 1981a. "Stages in the Empirical Programme of Relativism." *Social Studies of Science* 11:3–10.

Collins, H. M. 1981b. "The Place of the Core-Set in Modern Science: Social Contingency with Methodological Propriety in Science." *History of Science* 19:6–19.

Collins, H. M. 1981c. "What Is TRASP? The Radical Programme as a Methodological Imperative." *Philosophy of the Social Sciences* 11:215–224.

Collins, H. M. 1981d. "Son of Seven Sexes: The Social Destruction of a Physical Phenomenon." *Social Studies of Science* 11:33–62.

Collins, H. M., and T. J. Pinch. 1982. *Frames of Meaning: The Social Construction of Extraordinary Science*. London: Routledge and Kegan Paul.

Collins, H. M. 1983. "An Empirical Relativist Programme in the Sociology of Scientific Knowledge." In *Science Observed*, edited by K. Knorr-Cetina and M. Mulkay, pp. 85–114. London: Sage.

Collins, H. M. 1985. *Changing Order: Replication and Induction in Scientific Practice*. London: Sage.

Collins, H. M. 1988. "Public Experiments and Displays of Virtuosity: The Core-Set Revisited." *Social Studies of Science* 18:725–748 .

Collins, H. M. 1989. "The Meaning of Experiment: Replication and Reasonableness." In *Dismantling Truth: Reality in the Post-Modern World*, edited by H. Lawson and L. Appignanesi, pp. 82–92. London: Weidenfeld and Nicolson.

Collins, H. M. 1991. "Captives and Victims: Comment on Scott, Richards, and Martin." *Science, Technology and Human Values* 16:249–251.

Collins, H. M., and S. Yearley. 1992. "Epistemological Chicken." In *Science as Practice and Culture*, edited by A. Pickering, pp. 301–326. Chicago: University of Chicago Press.

Collins, H. M., and T. J. Pinch. 1993. *The Golem: What Everyone Should Know about Science*. Cambridge: Cambridge University Press.

Collins, H. M. 1994. "A Strong Confirmation of the Experimenter's Regress." *Studies in History and Philosophy of Science* 25:493–503.

Collins, H. M. 1999. "Tantalus and the Aliens: Publications, Audiences and the Search for Gravitational Waves." *Social Studies of Science* 29:163–197.

Collins, H. M. 2000. "Surviving Closure: Post-Rejection Adaptation and Plurality in Science." *American Sociological Review* 65:824–845.

Conant, J. B. 1951. "Pasteur's and Tyndall's Study of Spontaneous Generation." In *Harvard Case Histories in Experimental Science*, edited by J. B. Conant and L. K. Nash, vol. 1, pp. 11–61. Cambridge: Harvard University Press.

Cookson, C. 1989. "Test Tube Nuclear Fusion Claimed." *Financial Times*, March 23, A1, 28.

Crease, R. P., and N. P. Samios. 1989. "Cold Fusion Confusion." *New York Times Sunday Magazine*, September 24, 35–38.

Curtis, R. 1994. "Narrative Form and Narrative Force: Baconian Story-Telling in Popular Science." *Social Studies of Science* 24:419–461.

Cussins, C. 1996. "Ontological Choreography: Agency through Objectification in Infertility Clinics." *Social Studies of Science* 26:575–610.

Derrida, J. 1994. *Specters of Marx : The State of the Debt, the Work of Mourning, and the New International*. New York: Routledge.

Dewdney, A. K. 1997. *Yes, We Have No Neutrons: An Eye-Opening Tour through the Twists and Turns of Bad Science*. New York: Wiley.

Dolby, R.G. . 1979. "Reflections of Deviant Science." In *On the Margins of Science: The Social Construction of Rejected Knowledge*, edited by R. Wallis, pp. 9–48. Keele, U.K.: University of Keele Press.

Energy Research Advisory Board. 1989. *Cold Fusion Research*. U.S. Department of Energy, DOE/S0073.

Englehardt, H. and Caplan, A. 1987. *Scientific Controversies: Case Studies in the Resolution and Closure of Disputes in Science and Technology*. Cambridge: Cambridge University Press.

Epstein, S. 1993. *Impure Science: Aids, Activism, and the Politics of Knowledge*. Berkeley: University of California Press.

Evans-Pritchard, E. 1937. *Witchcraft, Oracles and Magic among the Azande*. Oxford: Clarendon Press.

Fadlon, J., and N. Lewin-Epstein. 1997. "Laughter Spreads: Another Perspective on Boundary Crossing in the Benveniste Affair." *Social Studies of Science* 27:131–141.
Farley, J., and G. L. Geison. 1974. "Science, Politics and Spontaneous Generation in Nineteenth-Century France: The Pasteur-Pouchet Debate." *Bulletin of the History of Medicine* 48:161–198.
Felt, U., and H. Nowotny. 1992. "Striking Gold in the 1990's: The Discovery of High-Temperature Superconductivity and Its Impact on the Science System." *Science, Technology & Human Values* 17:506–531.
Festinger, L., H. Riecken, and S. Schachter. 1956. *When Prophecy Fails: A Social and Psychological Study of a Modern Group That Predicted the Destruction of the World.* New York: Harper and Row.
Finn, R. 1989. "Caltech Finds No Evidence for Fusion in Attempts to Replicate Experiments." Press release; California Institute of Technology, May 1.
Fleischmann, M., S. Pons, and M. Hawkins. 1989a. "Electrochemically Induced Nuclear Fusion of Deuterium." *Journal of Electroanalytic Chemistry* 261: 301–308.
Fleischmann, M., S. Pons, and M. Hawkins. 1989b. "Erratum." *Journal of Electroanalytic Chemistry* 263:187.
Fleischmann, M., S. Pons, and R. Hoffman. 1989. "Measurements of Gamma-Rays from Cold Fusion." *Nature* 339:667.
Fleischmann, M., S. Pons, M. W. Anderson, L. J. Li, and M. Hawkins. 1990. "Calorimetry of the Palladium-Deuterium Heavy Water System." *Journal of Electroanalytical Chemistry* 287:293–320.
Fleischmann, M., and S. Pons. 1993. "Calorimetry of the Pd=$D_2O$.System: From Simplicity via Complications to Simplicity." *Physics Letters* A 176:118–129.
Fleischmann, M., and S. Pons. 1994a. "Heat after Death." *Proceedings of the 4th International Conference on Cold Fusion*, vol. 2, pp. 87–95. Stanford, Calif.: Electric Power Research Institute.
Fleischmann, M., and S. Pons. 1994b. "Reply to the Critique of Morrison." *Physics Letters* A 187:276–280.
Fox, H. 1992. *Cold Fusion Impact in the Enhanced Energy Age*. Salt Lake City: Fusion Information Center.
Franklin, A. 1994. "How to Avoid the Experimenter's Regress." *Studies in the History and Philosophy of Science* 25:463–491.
Franks, F. 1982. *Polywater*. Cambridge: MIT Press.
Freud, S. 1958. "The 'Uncanny.'" In *On Creativity and the Unconcious*, edited by B. Nelson, pp. 122–161. New York: Harper and Row.
Friedlander, M. 1995. *At the Fringes of Science*. Boulder: Westview Press.
Fujimura, J. 1987. "Constructing 'Do-Able' Problems in Cancer Research: Articulating Alignment." *Social Studies of Science* 17:257–293.
Fujimura, J. 1992. "Crafting Science: Standardized Packages, Boundary Objects, and Translation." In *Science as Practice and Culture*, edited by A. Pickering, pp. 168–214. Chicago: University of Chicago Press.
Galison, P. 1987. *How Experiments End*. Chicago: University of Chicago Press.
Gamson, J. 1997. "Messages of Exclusion: Gender, Movements, and Symbolic Boundaries." *Gender and Society* 11:178–199.
Gellner, E. 1970. "Concepts and Society." In *Rationality*, edited by B. Wilson, pp. 18–46. Oxford: Basil Blackwell.
Gieryn, T. F. 1983. "Boundary-Work and the Demarcation of Science from Non-Science: Strains and Interests in Professional Ideologies of Scientists." *American Sociological Review* 48:781–795.
Gieryn, T. F. 1992. "The Ballad of Pons and Fleischmann: Experiment and Narrative

in the (Un)Making of Cold Fusion." In *The Social Dimensions of Science*, edited by E. McMullin, pp. 272–294. Notre Dame, Ind.: University of Notre Dame Press.

Gieryn, T. F. 1995. "Boundaries of Science." In *Handbook of Science and Technology Studies*, edited by S. Jasanoff and T. J. Pinch, pp. 393–443. Thousand Oaks, Calif.: Sage.

Gieryn, T. F. 1999. *Cultural Boundaries of Science: Credibility on the Line*. Chicago: University of Chicago Press.

Goffman, E. 1969. *The Presentation of Self in Everyday Life*. London: Allen Lane, Penguin Press.

Gooding, D., T. Pinch, and S. Schaffer. 1989. *The Uses of Experiment: Studies in the Natural Sciences*. Cambridge: Cambridge University Press.

Goodstein, D. 1994. "Pariah Science: What Happened to Cold Fusion." *American Scholar* 63:527–552.

Grayson, L. 1995. *Scientific Deception: An Overview and Guide to the Literature of Misconduct and Fraud in Scientific Research*. London: British Library.

Griggs, J. L. 1994. "A Brief Introduction to the Hydrosonic Pump and the Associated 'Excess Energy' Phenomenon." Proceedings of the 4th International Conference on Cold Fusion, pp. 43–53. Stanford, Calif.: Electric Power Research Institute.

Gross, A. 1995. "Renewing Neo-Aristotelian Theory: The Cold Fusion Controversy as a Test Case." *Quarterly Journal of Speech* 81:48–62.

Gryzinski, M. 1989. Letter to the Editor. *Nature* 338:712.

Hacking, I. 1983. *Representing and Intervening*. Cambridge: Cambridge University Press.

Hacking, I. 1993. "Historical Epistemology." Paper presented at the Workshop on Historical Epistemology, at the University of Toronto.

Hacking, I. 1999. *The Social Construction of What?* Cambridge: Harvard University Press.

Haraway, D. 1992. "The Promises of Monsters: A Regenerative Politics for Inappropriate/d Others." In *Cultural Studies*, edited by L. Grossberg, C. Nelson, and P. Treichler. New York: Routledge.

Haraway, D. 1994. "A Game of Cat's Cradle: Science Studies, Feminist Theory, Cultural Studies." *Configurations: A Journal of Literature and Science* 1:59–71.

Hard, M. 1994. "Technology as Practice: Local and Global Closure Processes in Diesel-Engine Design." *Social Studies of Science* 24:549–585.

Hazen, R. M. 1988. *The Breakthrough: The Race for the Superconductor*. New York: Simon and Schuster.

Hess, D. 1992. "Disciplining Heterodoxy, Circumventing Discipline: Parapsychology, Anthropologically." *Knowledge and Society* 9:223–252.

Heylin, M. 1989. Editorial. *Chemistry and Engineering News*, April 3, 3.

Hilgartner, S. 1990. "The Dominant View of Popularization: Conceptual Problems, Political Uses." *Social Studies of Science* 20:519–539.

Hoffman, N. 1995. *A Dialogue on Chemically Induced Nuclear Effects: A Guide for the Perplexed about Cold Fusion*. La Grange Park, Ill.: American Nuclear Society.

Huizenga, J. R. 1992. *Cold Fusion: The Scientific Fiasco of the Century*. New York: Oxford University Press.

Indyk, D., and D. Rier. 1993. "Grassroots Aids Knowledge." *Knowledge: Creation, Diffusion, Utilization* 15:3–43.

ISI. 1990a. "The Hottest Fields of 1989." *Science Watch* 1, 3:1.

ISI. 1990b. "Scientists Vote on Cold Fusion: Their Verdict? No, Not Likely." *Science Watch*, 1, 2:7.

Johnson, K. H., and D. P. Clougherty. 1989. "Hydrogen-Hydrogen/Deuterium-Deuterium Bonding in Palladium and the Superconducting/Electrochemical Properties of $PdH_x/PdD_x$." *Physics Letters B* 3:795–803.

Jones, S. E., et al. 1989. "Observation of Cold Nuclear Fusion in Condensed Matter." *Nature* 338:737–740.

Jones, S. E., and L. D. Hansen. 1995. "Examination of Claims of Miles et al. in Pons-Fleischman-Type Cold Fusion Experiments." *Journal of Physical Chemistry B* 99:6973–6979.

Jones, S. E., L. D. Hansen, and D. S. Shelton. 1998. "Discussion of Reply to 'Examination of Claims of Miles et al. in Pons-Fleischmann-Type Cold Fusion Experiments.'" *Journal of Physical Chemistry* B 102:3644–3648.

Kevles, D. 1993. *Cold Facts. New Yorker*, August 2, 82.

Kim, K-M. 1996. "Hierarchy of Scientific Consensus and the Flow of Dissensus Over Time." *Philosophy of the Social Sciences* 26:3–25.

Klotz, I., and J. Katz. 1991. "Extraordinary Electrical Experiment: Two of Them." *American Scholar* 60:247–250.

Knorr-Cetina, K. 1979. "Tinkering toward Success: Prelude to a Theory of Scientific Practice." *Theory and Society* 8:347–376.

Knorr-Cetina, K. D. 1981. *The Manufacture of Knowledge: An Essay on the Constructivist and Contextual Nature of Science*. Oxford: Pergamon Press.

Kohn, A. 1988. *False Prophets: Fraud and Error in Science and Medicine*. Oxford: Basil Blackwell.

Kolata, G. 1994. "Viruses or Prions: An Old Debate Still Rages." *New York Times*, October 4, C1.

Koonin, S., and M. Nauenberg. 1989. "Calculated Fusion Rates in Isotopic Hydrogen Molecules." *Nature* 339:690–692.

Kuhn, T. S. 1970. *The Structure of Scientific Revolutions*. 2nd edition. Chicago: University of Chicago Press.

Langmuir, I. 1989. "Pathological Science," edited by R. N. Hall. *Physics Today* 42 (October):36–48.

Latour, B., and S. Woolgar. 1979. *Laboratory Life: The [Social] Construction of Scientific Facts*. Princeton: Princeton University Press.

Latour, B. 1981. "Insiders and Outsiders in the Sociology of Science; or How Can We Foster Agnosticism." *Knowledge and Society* 3:199–216.

Latour, B. 1983. "Give Me a Laboratory and I Will Raise the World." In *Science Observed: Perspectives on the Social Study of Science*, edited by K. Knorr-Cetina and M. Mulkay. London: Sage.

Latour, B. 1986. "The Powers of Association." In *Power, Action and Belief: A New Sociology of Knowledge?* edited by J. Law, pp. 264–280. London: Routledge and Kegan Paul.

Latour, B. and Bastide, F. 1986. "Writing Science—Fact and Fiction: The Analysis of the Process of Reality Construction through the Application of Socio-Semiotic Methods to Scientific Texts." In *Mapping the Dynamics of Science and Technology*, edited by M. Callon, J. Law, and A. Rip, pp. 51–66. London: Macmillan.

Latour, B. 1987. *Science in Action: How to Follow Scientists and Engineers through Society*. Cambridge: Harvard University Press.

Latour, B. 1988a. *The Pasteurization of France*. Cambridge: Harvard University Press.

Latour, B. 1988b. "'The Prince' for Machines as Well as for Machinations." In *Technology and Social Change*, edited by B. Elliot, pp. 304–317. Edinburgh: Edinburgh University Press.

Latour, B. 1989. "Clothing the Naked Truth." In *Dismantling Truth: Reality in the Post-Modern World*, edited by H. Lawson and L. Appignanesi, pp. 101–126. London: Weidenfeld and Nicolson.

Latour, B. 1990. "The Force and Reason of Experiment." In *Experimental Enquiries*, edited by H. LeGrand. Dordrecht: Reidel.

Latour, B., P. Mauguin, and G. Teil. 1992. "A Note on Socio-Technical Graphs." *Social Studies of Science* 22:33–57.

Latour, B. 1993. *We Have Never Been Modern*. Cambridge: Harvard University Press.
Latour, B. 1996. "Do Scientific Objects Have a History? Pasteur and Whitehead in a Bath of Lactic Acid." *Common Knowledge* 5:76–91.
Latour, B. 1999. "For David Bloor . . . and Beyond. A Reply to David Bloor's Anti-Latour." *Studies in the History and Philosophy of Science* 30:113–130.
Laudan, L. 1977. *Progress and Its Problems: Toward a Theory of Scientific Growth*. Berkeley: University of California Press.
Laudan, L. 1983. "The Demise of the Demarcation Problem." In *Physics, Philosophy and Psychoanalysis*, edited by R. S. Cohen and L. Laudan, pp. 73–80. Dordrecht: D. Reidel.
Law, J. 1986. "The Heterogeneity of Texts." In *Mapping the Dynamics of Science and Technology*, edited by M. Callon, J. Law, and A. Rip, pp. 67–83. London: Macmillan.
Lawrence Livermore National Laboratory. 1990. "Roundtable Discussion on Cold Fusion." *Energy and Technology Review*, October, 21–23.
LeGrand, H. E. 1988. *Drifting Continents and Shifting Theories*. Cambridge: Cambridge University Press.
Lewenstein, B. 1991. "Preserving Data about the Knowledge Creation Process." *Knowledge: Creation, Diffusion, Utilization* 13:79–86.
Lewenstein, B. 1992a. "Cold Fusion and Hot History." *Osiris* 7:135–163.
Lewenstein, B. 1992b. "Cold Fusion Saga: Lesson in Science." *Forum for Applied Research and Public Policy* 7:67–77.
Lewenstein, B. 1995. "From Fax to Facts: Communication in the Cold Fusion Saga." *Social Studies of Science* 25:403–436.
Lewis, N., et al. 1989. "Searches for Low-Temperature Nuclear Fusion of Deuterium in Palladium." *Nature* 340:525–528.
Lievrouw, L. 1990. "Communication and the Social Representation of Scientific Knowledge." *Critical Studies in Mass Communication* 7:1–10.
Lindley, D. 1989. "Cold Fusion: Still No Certainty." *Nature* 339:84.
Lindley, D. 1990. "The Embarrassment of Cold Fusion." *Nature* 344:375.
Mackle, G. and J. Peterson. 1981. "Controversies in Science and Technology: A Protocol for Comparative Research." *Science, Technology and Human Values* 6:6–15.
Mallove, G. 1991. *Fire from Ice: Searching for the Truth Behind the Cold Fusion Furor*. New York: Wiley.
Mallove, E. 1994. "Texas A&M Professor Attacked for Cold Fusion and 'Alchemy.'" *Cold Fusion* 1:21–22.
Mallove, E. 1995. "The Tip of an Iceberg." *Infinite Energy* 1:1.
Mallove, E. 2000. "The Triumph of Alchemy: Professor John Bockris and the Transmutation Crisis at Texas A&M." *Infinite Energy* 6:32.
Manning, J. 1996. *The Coming Energy Revolution*. New York: Avery Publishing Group.
Martin, B. 1992. "Scientific Fraud and the Power Structure of Science." *Prometheus* 10:83–98.
McAllister, J. 1992. "Competition among Scientific Disciplines in Cold Nuclear Fusion Research." *Science in Context* 5:17–50.
McKinney, W. 1998. "Where Experiments Fail: Is 'Cold Fusion' Science as Normal?" In *A House Built on Sand: Exposing Postmodernist Myths About Science*, edited by N. Koertge, pp. 133–150. Oxford: Oxford University Press.
Merton, R. K. 1973. "The Normative Structure of Science." In *The Sociology of Science: Theoretical and Empirical Investigations*, edited by N. W. Storer, pp. 254–278. Chicago: University of Chicago Press.
Michael, M., and L. Birke. 1994. "Enrolling the Core-Set: The Case of the Animal Experimentation Controversy." *Social Studies of Science* 24:81–95.
Miles, M. H. 1998. "Reply to 'Examination of Claims of Miles et al. in Pons-Fleischmann-Type Cold Fusion Experiments.'" *Journal of Physical Chemistry B* 102:3642–3644.

Miley, G. 2000. "Some Personal Reflections on Scientific Ethics and the Cold Fusion 'Episode.'" *Accountability in Research* 8:121–126.
Miley, G., and J. Patterson. 1996. "Nuclear Transmutations in Thin-Film Nickel Coatings Undergoing Electrolysis." Presented at Second Annual Conference on Low Energy Nuclear Reactions, College Station, Tex., Sept. 13–14.
Mills, R. L., and S. P. Kneizys. 1991. "Excess Heat Production by the Electrolysis of an Aqueous Potassium Carbonate Electrolyte and the Implications for Cold Fusion." *Fusion Technology* 20:65–78.
Mizuno, T. 1998. *Nuclear Transmutation: The Reality of Cold Fusion*. Concord, N.H.: Infinite Energy Press.
Morrison, D. 1990. "The Rise and Decline of Cold Fusion." *Physics World* 35 (February):35–39.
Morrison, D. 1994. "Comments on Claims of Excess Enthalpy by Fleischmann and Pons Using Simple Cells Made to Boil." *Physics Letters* A 185:498–502.
Mukerji, C., and B. Simon. 1998. "Out of the Limelight: Discredited Communities and Informal Communication on the Internet." *Sociological Inquiry* 68:258–273.
Mulkay, M. J. 1976. "Norms and Ideology in Science." *Social Science Information* 15:627–656.
Mulkay, M. J. 1979. *Science and the Sociology of Knowledge*. London: George Allen & Unwin.
Mulkay, M. and G. N. Gilbert. 1986. "Replication and Mere Replication." *Philosophy of Social Science* 16:21–37.
Nelkin, D. 1979. *Controversy: Politics of Technical Decisions*. Beverly Hills, Calif.: Sage.
Nelkin, D. 1996. "The Science Wars: Responses to a Marriage Failed." In *Science Wars*, edited by A. Ross, pp. 114–122. Durham: Duke University Press.
Noble, D. 1992. *A World without Women: The Christian Clerical Culture of Western Science*. New York: Knopf.
Notoya, R. 1993 "Cold Fusion by Electrolysis in a Light Water-Potassium Carbonate Solution with a Nickel Electrode." *Fusion Technology* 24:202–210.
Nowotny, H., and H. Rose, eds. 1979. *Counter-Movements in the Sciences*. Boston: D. Reidel.
Nye, M. J. 1980. "N-Rays: An Episode in the History and Psychology of Science." *Historical Studies in the Physical Sciences* 6:125–156.
Oriani, R. A. 1993. "Cold Fusion Difficulty." *Science* 261:279.
Packer, K., and A. Webster 1996. "Patenting Culture in Science: Reinventing the Scientific Wheel of Credibility." *Science, Technology and Human Values* 21:427–453.
Park, R. 2000. *Voodoo Science: The Road from Foolishness to Fraud*. Oxford: Oxford University Press.
Peat, F. D. 1989. *Cold Fusion: The Making of a Scientific Controversy*. Chicago: Contemporary Books.
Petrasso, R., et al. 1989. "Problems with the Gamma-Ray Spectrum in the Fleischmann et al. Experiments." *Nature* 339:183–185.
Petrasso, R. "Reply." *Nature* 339:667–669.
Picart, C. 1994. "Scientific Controversy as Farce: The Benveniste-Maddox Counter Trials." *Social Studies of Science* 24:7–37.
Pickering, A. 1980. "The Role of Interests in High-Energy Physics: The Choice between Charm and Colour." In *The Social Process of Scientific Investigation*, edited by K. Knorr, R. Krohn, and R. Whitley, pp. 139–178. Dordrecht: Reidel.
Pickering, A. 1984. *Constructing Quarks: A Sociological History of Particle Physics*. Chicago: University of Chicago Press.
Pickering, A. 1989. "Editing and Epistemology: Three Accounts of the Discovery of the Weak Neutral Current." *Knowledge and Society: Studies in the Sociology of Science* 8:217–232.

Pickering, A., ed. 1992. *Science as Practice and Culture*. Chicago: University of Chicago Press.
Pickering, A. 1995. *The Mangle of Practice: Time, Agency and Science*. Chicago: University of Chicago Press.
Pinch, T. J. 1985. "Towards an Analysis of Scientific Observation: The Externality and Evidential Significance of Observational Reports in Physics." *Social Studies of Science* 15:3–36.
Pinch, T. J. 1986a. *Confronting Nature: The Sociology of Solar-Neutrino Detection*. Dordrecht: D. Reidel.
Pinch, T. J. 1986b. "Strata Various." *Social Studies of Science* 16:705–713.
Pinch, T. J. 1994. "Cold Fusion and the Sociology of Scientific Knowledge." *Technical Communication Quarterly* 3:85–102.
Pinch, T. J. 1996. "Rhetoric and the Cold Fusion Controversy: From the Chemists' Woodstock to the Physicists' Altamont." In *Science, Reason, and Rhetoric*, edited by H. Krips, J. E. McGuire, and T. Melia, pp. 153–176. Pittsburgh: University of Pittsburgh Press.
Pinch, T. J. 1999. "Half a House: A Response to McKinney." *Social Studies of Science* 29:235–240.
Pool, R. 1989a. "In Hot Water over Cold Fusion." *Science* 156:500.
Pool, R. 1989b. "Skepticism Grows over Cold Fusion." *Science* 244:285.
Prelli, L. 1989. "The Rhetorical Construction of Scientific Ethos." In *Rhetoric in the Human Sciences*, edited by H. Simons, pp. 87–104. London: Sage.
Randi, J. 1994. "Waves and Vibrations." *APS News*, June, 4.
Rousseau, D. 1992. "Case Studies in Pathological Science." *American Scientist* 80:54–62.
Rudwick, M. 1985. *The Great Devonian Controversy: The Shaping of Scientific Knowledge among Gentlemanly Specialists*. Chicago: University of Chicago Press.
Salamon, R., et al. 1990. "Limits on the Emission of Neutrons and Gamma-Rays, Electrons and Protons from Pons/Fleischmann Electrolytic Cells." *Nature* 344:401–405.
Schiff, M. 1994. *The Memory of Water: Homeopathy and the Battle of Ideas in the New Science*. San Francisco: Thorsons.
Scott, P., E. Richards, and B. Martin. 1990. "Captives of Controversy: The Myth of the Neutral Social Researcher in Contemporary Scientific Controversies." *Science, Technology and Human Values* 15:474–494.
Secord, J. A. 1989. "Extraordinary Experiment: Electricity and the Creation of Life in Victorian England." In *The Uses of Experiment: Studies in the Natural Sciences*, edited by D. Gooding, T. Pinch, and S. Schaffer, pp. 337–383. Cambridge: Cambridge University Press.
Shapin, S. 1984. "Pump and Circumstance: Robert Boyle's Literary Technology." *Social Studies of Science* 14:481–520.
Shapin, S., and S. Schaffer. 1985. *Leviathan and the Air Pump: Hobbes, Boyle and the Experimental Life*. Princeton: Princeton University Press.
Shapin, S. 1993. *Trust, Honesty, and the Credibility of Science*. Committee on Social and Ethical Impacts of Advances in Biomedicine, NAS.
Shapin, S. 1994. *A Social History of Truth: Civility and Science in Seventeenth-Century England*. Chicago: University of Chicago Press.
Simon, B. 1999. "Undead Science: Making Sense of Cold Fusion after the (Arti)Fact." *Social Studies of Science* 29:61–85.
Simon, B. 2001. "Public Science: Media Configuration and Closure in the Cold Fusion Controversy." *Public Understanding of Science* 10:1–20.
Sismondo, S. 1996. *Science without Myth: On Constructions, Reality, and Social Knowledge*. Albany: State University of New York Press.
Smith, D. 1989. "Quest for Fusion." *CalTech Engineering and Science*, Summer, 6–19.

Sokal, A. 1996. "A Physicist Experiments with Cultural Studies." *Lingua Franca* 6:62–64.
Sproull, L., and S. Kiesler. 1991. *Connections: New Ways of Working in the Networked Organization*. Cambridge: MIT Press.
Srinivasan, M. 1994. "Technical Note." *Cold Fusion* 1:73.
Star, S. L., and J. Griesemer. 1989. "Institutional Ecology, 'Translations' and Boundary Objects: Amateurs and Professionals in Berkeley's Museum of Vertebrate Zoology, 1907–1939." *Social Studies of Science* 19:387–420.
Stipp, D. 1989. "Groups of Physicists, Releasing Reams of Data, Dispute Claims of Cold Fusion." *Wall Street Journal*, May 2, B3.
Stirling, S. M. 1991. "Roachstompers." In *Power*, edited by S. M. Stirling, pp. 223–258. New York: Baen Books.
Storms, E. 1991. "Review of Experimental Observations about the Cold Fusion Effect." *Fusion Technology* 20:433–442.
Storms, E. 1993. "The Status of Cold Fusion as a Significant Phenomenon." *21st Century Science and Technology*. July:84–88.
Storms, E. 1994. "Warming up to Cold Fusion." *Technology Review*, June, 19–29.
Storms, E. 1996a. "A Critical Review of the 'Cold Fusion' Effect." *Journal of Scientific Exploration* 10:185–192.
Storms, E. 1996b. "How to Produce the Pons-Fleischmann Effect." *Journal of Fusion Technology* 29:261.
Storms, E. 2001. "Whatever Happened to Cold Fusion." Unpublished Manuscript.
Strauss, A. 1978. "A Social Worlds Perspective." *Studies of Symbolic Interaction* 1:119–128.
Sullivan, D. L. 1994. "Exclusionary Epideictic: Nova's Narrative Excommunication of Fleischmann and Pons." *Science, Technology and Human Values* 19:283–306.
Taubes, G. 1990. "Cold Fusion Conundrum at Texas A&M." *Science* 248:1299–1304.
Taubes, G. 1993. *Bad Science: The Short Life and Weird Times of Cold Fusion*. New York: Random House.
Taylor, C. A. 1996. *Defining Science: A Rhetoric of Demarcation*. Madison: University of Wisconsin Press.
Thomsen, M. 1998. "Learning Lessons about Ethics." *Physics World* 12(March):3.
Tinsley, C. 1997. "An Interview with Martin Fleischmann." *Infinite Energy* 3:13–14,66.
Toumey, C. 1996. "Conjuring Science in the Case of Cold Fusion." *Public Understanding of Science* 5:121–133.
Travis, G. D. L. 1981. "Replicating Replication? Aspects of the Social Construction of Learning in Planarian Worms." *Social Studies of Science* 11:11–32.
Traweek, S. 1988. *Beamtimes and Lifetimes: The World of High Energy Physics*. Cambridge: Harvard University Press.
Walling, C., and J. Simons. 1989. "Two Innocent Chemists Look at Cold Fusion." *Journal of Physical Chemistry* 93:4693–4696.
Wallis, R., and P. Morley. 1976. *Marginal Medicine*. London: Peter Owen.
Wallis, R., ed. 1979. *On the Margins of Science: The Social Construction of Rejected Knowledge*. Keele, U.K.: University of Keele Press.
Webster, A. J. 1979. "Scientific Controversy and Socio-Cognitive Metonymy: The Case of Acupuncture." In *On the Margins of Science: The Social Construction of Rejected Knowledge*, edited by R. Wallis, pp. 121–138. Keele, U.K.: University of Keele Press.
Westfall, R. 1980. "The Influence of Alchemy on Newton." In *Science, Pseudo-Science and Society*, edited by M. Hanen, M. Osler, and R. Weyant, pp. 145–169. Waterloo, Ont.: Wilfrid Laurier Press.
White, N. 1993. "The Fleischmann-Pons Cell Boiloff: Is It Moonshine?" *21st Century Science and Technology* 6(Sept. 1):70–74.
Williams, D. 1993. Book Review. *Physics Today* 46(January):74.
Wolpert, L. 1992. *The Unnatural Nature of Science*. London: Faber & Faber.

# Index

# *About the Author*

Bart Simon is an assistant professor in the Department of Sociology and Anthropology at Concordia University in Montreal, Canada. He completed his Ph.D. on the sociology of the cold fusion controversy in 1999 and has since started research on Theosophy and science in late-nineteenth-century India, surveillance and information technologies, and the sociomateriality of computer games. He currently teaches courses in the sociology of cyberspace, material culture, and science and technology studies.